The Ultimate *Roy Rogers* Collection

Identification & Price Guide

Ron Lenius

Published by

Please call or write for our free catalog of publications. Our toll-free number to place an order or obtain a free catalog is 800-258-0929 or please use our regular business telephone 715-445-2214.

Library of Congress Catalog Number 2001088594
ISBN 0-87349-226-9

DEDICATED
TO

ROY ROGERS
"The King of the Cowboys"

AND

DALE EVANS
"The Queen of the West"

SELECTED BIBLIOGRAPHY

Phillips, Robert W. *Roy Rogers*. Jefferson, North Carolina: McFarland & Company, Inc., Publishers, 1995.

Rogers, Jr., Roy, with Karen Ann Wohahn. *Growing Up with Roy and Dale*. Ventura, California: Regal Books, 1986.

Rothel, David. *The Roy Rogers Book*. Madison, North Carolina: Empire Publishing, Inc. 1987.

Personal Interview with Roy Rogers, Jr., audio recorded. Victorville, California, October, 2000.

CONTENTS

INTRODUCTION

I was introduced to the "King of the Cowboys," Roy Rogers, and his movie co-stars on the silver screen of the Door Theatre in Sturgeon Bay, Wisconsin. It was in the "good old days" of the early fifties when Roy, Dale Evans, "Queen of the West," and their pals rode into my young heart at a Saturday afternoon matinee. When the doors opened for the show, the other kids and I would run down the aisle hoping to grab a front row center seat to see our hero and his friends in an action-packed western adventure. After the lights went down and the show began, I would sit spellbound on the edge of my seat watching Roy astride Trigger, "the smartest horse in the movies," bring the bad guys to justice and save the day. A round of applause and cheers from an audience of "buckaroos" ended another great Roy Rogers and Dale Evans movie.

I and a lot of other boys fantasized about being Roy Rogers when we played cowboys and Indians with our friends, or when we were alone day-dreaming. One year for Christmas my dad gave a beautiful Appaloosa horse to me and my oldest sister, Nancy. It was not a smart horse like Trigger, whom Roy said could turn on a dime and give nine cents change. Our horse was on the wild side and loved to run. One time when Nancy was riding him gently around the barnyard, he bolted out of the yard and went running as fast as he could down the highway with her hanging on for dear life. Well, my dad piled the rest of the family in the car and we went tearing after them, finally stopping them several miles away–just like in a Roy Rogers movie. I never did get a chance to ride him after this. The last straw for this horse came when he bit my mom in the seat of the pants. He ended up being sold to a horse racing enterprise and winning a lot of races.

Since I was now horseless, I named my favorite cow "Nellybelle" and rode her when bringing the herd of cows back to the barn from pasture. The name suited her well as she wore a big old cowbell and acted a lot like Pat Brady's ornery jeep. I still remember the shocks I got from her brushing my legs against the electric fence while riding down the narrow fenced lane to fetch the other cows for milking.

When I was in grade school, I acquired my first Roy Rogers guitar by selling garden seeds from American Seed Company to the neighboring farms. After school when my chores were done, I would take this little orange toy guitar, climb up into the silo (which was like an echo chamber) and serenade the cows in the barn with my music. Unfortunately, my dad thought that the cows gave less milk because of my singing and ordered me to practice away from the barn.

After a lot of "I wanna" persuasion from me and my sister, my folks finally purchased our first television set. We were probably the last people in Door County to get one. But, at last, our whole family got to sit down and watch "The Roy Rogers Show" every Sunday evening. Between 1951 and 1955 over a hundred episodes of this half-hour show were produced. General Foods, producers of Post brand cereals, sponsored "The Roy Rogers Show" on radio and television; and sold about two and a half billion boxes of cereal with Roy's picture or likeness printed on them. Roy Rogers trading cards and other premiums were inside the boxes. The back of the box advertised photos, pin-backs, and other Roy Rogers items that you could receive by sending in the box top and a quarter.

Through radio, movies, television, magazines, books, and live performances, millions of people discovered Roy Rogers, the greatest American hero for all age groups. It was estimated that more than eighty million people in the United States and Canada saw Roy Rogers' movies in 1948. This number greatly increased through the years with the production of more Roy Rogers and Dale Evans films. From 1943 to 1954, Roy was the number one western box office star. Their movies and television shows not only entertained us with music, drama, and gripping action, but also provided us with inspirational family values. We were all touched by the inspiring performances, and acts of love, kindness, and charity portrayed by Roy and Dale on the screen and in their personal life.

During the 1950s, Roy had over 2000 fan clubs in the United States. He also had the largest single fan club in the world located in London, England, with fifty thousand members. In 1947, he was voted the most popular movie star in England. Between 1945 and 1975, Roy Rogers and Dale Evans held more all-time box office records at state fairs, rodeos, and other personal appearances than anyone else in the entertainment profession. Roy had starred in 87 musical westerns for Republic Pictures and in 13 other motion picture films for other studios.

The *Saturday Evening Post, Life, Look, McCalls, Good Housekeeping,* and other "family" magazines wrote feature stories about Roy and Dale's family life. Their lives were filled to the brim with mountains of happiness, valleys of grief, and hectic Hollywood schedules. The nation and the world felt their joy when they adopted six of their nine children, and we wept for them when three of their children passed away. Roy and Dale were sustained by their iron-clad faith and great love of God and each other.

Presidents, heads of state, and others have bestowed many awards and honors upon Roy Rogers and Dale Evans for their vast amount of time and resources given to various charities, orphanages, and other organizations in need of their services. Roy is the only performer who has been elected twice into the Country Music Hall of Fame, and was also inducted into the National Hall of Fame. He won two Golden Globe Awards, and was honored with four stars in the Hollywood Walk of Fame for his work in radio, records, motion pictures, and television. Roy and Dale gave many free concerts for worthy causes, including performances for the servicemen in World War II and Vietnam. Sufficient praise cannot be given to this incredible couple who gave so much of themselves to so many. During their fifty-plus years of marriage, they practiced what they preached in their Christian faith, love of humanity, and family values.

We mourn the loss of Roy Rogers and Dale Evans-American institutions to generations of people around the world who were touched by their talent, charisma, and ideals. Roy was our greatest hero and a sweetheart of a man who warmed our souls with his true smile and the twinkle in his eye.

Roy's popularity was evidenced upon his passing by major newspapers, magazines, and television stations in our nation and in countries around the world honoring him with memorials on his life and the wonderful legacy he left us. Just one example is the code of conduct he prescribed for boys and girls which reads:

The Roy Rogers Riders Club Rules:

1. Be neat and clean.
2. Be courteous and polite.
3. Always obey your parents.
4. Protect the weak and help them.
5. Be brave, but never take chances.
6. Study hard and learn all that you can.
7. Be kind to animals and care for them.
8. Eat all your food and never waste any.
9. Love God and go to Sunday School regularly.
10. Always respect our flag and our country.

This really was great advice for all of us. Roy Rogers also sponsored a safety bike-riding program for school children.

Roy was not only the "King of the Cowboys," but had replaced Hopalong Cassidy as "King of the Merchandisers." Hopalong had shown Roy the golden road of merchandising, which he turned into a "happy trail" of marketing his name on quality items. Art Rush, Roy's manager, stated that between 1945 and 1975 consumers had spent more than one billion dollars on the name Roy Rogers. His well-recognized name appeared on over 400 licensed items between 1945 and 1955. He was second only to Walt Disney for the number of commercial manufacturing contracts he held. America became saturated with Roy Rogers, Dale Evans, and their co-stars memorabilia through radio, movies, records, television, magazines, books, food products, toys, etc., which was unprecedented.

Roy Rogers, "The Singing Cowboy," had more impact on merchandising his name than any other western movie star. Large department and catalog stores such as Montgomery Ward and Sears, Roebuck sold every variety of toys, clothes, and interior furnishings bearing Roy Rogers and Dale Evans' names. During the 1950s, more than 125 products were made by 74 different manufacturers. There were Roy Rogers beds, bed coverings, pillows, curtains, food products, clocks, jewelry, clothes, dishes, toys, rugs, cameras, and just about everything else that a child could wear, play with, and put in their rooms.

A vital part of the marketing appeal for parents was that all of the authorized items marketed bore Roy's personal seal of approval. This assured the consumer that they were purchasing a safe and quality-made product for a fair price. Roy "Dusty" Rogers, Jr., states that, "every product was tested on us (Roy and Dale's children) before it hit the public. So, if we beat the stuff to death, and it was still in good shape, it would go to market." As a testament to the durability of these items that passed Roy's testing program, an incredible amount of these collectibles have survived through the decades.

It has been my sincere privilege and pleasure to have photographed and listed with information and estimated values the most comprehensive listing of these rare and wonderful vintage collectibles and memorabilia that were produced from 1938 to 1965. I hope that you will experience the same joy, inspiration and excitement that I received upon viewing these items for the very first time. May you discover the ones you are looking for hidden away in your grandparent's attic, or at your neighbor's garage sale.

Happy Hunting and Happy Trails to you!

Ron Lenius

ACKNOWLEDGMENTS

I am tremendously indebted to many great people who have given me their time, energy, and wisdom in creating this book. Their wonderful enthusiasm, encouragement, and knowledge have provided the stimulation and information for me to complete this project. I sincerely thank them for their gracious efforts, and am deeply grateful for their generous support.

Many thanks to the following Roy Rogers collectors who have contributed their expertise on Roy and Dale collectibles and allowed me to use photos of their very special collections.

Roy "Dusty" Rogers, Jr., and the Roy Rogers and Dale Evans Museum
Chuck Quinn
Gary Schneekloth
Brian Wolf
Mike Moore
Larry Ciserwski
David Tims

I owe a very big thank you to Paul Kennedy, book acquisitions editor at Krause Publications for believing in me and my dream of getting this book published. Paul seriously went to bat for me many times. I was very close to giving up on this project when he gave me the news that made my dream come true. Also, a great deal of thanks to Karen O'Brien, associate editor of books. She is the editor of this project and my guiding light on the "Happy Trails" of book cover design. Thank you Jamie Martin for the book's beautiful design and thanks to all the departments at Krause Publications.

A special thank you to Mary Lyn Lucchese for enduring the pain of typing the manuscript and other items for me from almost day one on this project. Mary Lyn and her husband Mario have been a constant support team and a wonderful friendship. To both of you, thank you very much!

Mary Lyn Lucchese

I am very indebted to another great friend, Chuck Quinn, who started me on the trail of collecting Roy Rogers and other western hero toys. Chuck and his wife, Dorothy, have graciously allowed me to photograph his huge Roy Rogers and Dale Evans collection.

Chuck Quinn with Roy Rogers

Roy "Dusty" Rogers, Jr., has been most generous with his time, kindness, and wealth of information by consenting to an interview, and providing special photographs from the archives of the Roy Rogers and Dale Evans Museum. Thank you Dusty! And thanks to your family.

To my best friend and wife, Roxanne, I give my greatest heartfelt thanks for her bountiful love, encouragement, and support. This book could not have been completed without her enormous help.

Roxanne Lenius

My love and thanks to my parents, Eileen and Lloyd, for teaching me the greatest values in life.

INTERVIEW WITH ROY "DUSTY" ROGERS, JR.

Author Ron Lenius: I know that you are a collector of Roy Rogers and Dale Evans memorabilia and collectibles. Do you collect other things as well?

Dusty: Yes, if you look at my office here, I have a little bit of everything. When I was a kid, one of the things that I loved to play with was the little six-inch Marx cowboys and Indians, as well as army men. I had tons of them. Then you could buy them for ten cents apiece. The last one I bought cost me twenty-five dollars. My brother, Sandy, and I played cowboys and Indians and army a lot. Sandy loved army and the Civil War stuff. We used to use our Red Ryder BB guns to shoot the plastic army men.

Roy "Dusty" Rogers, Jr.

I collect anything that I really like; this includes a Lone Ranger silver bullet, a Tom Mix badge, and the Roy and Dale stuff.

A lot of people would ask me who was my hero when I was young. Well, of course, my Dad, along with the Lone Ranger, Clayton Moore. He was the good guy behind the mask. There was a mystique about him, like Zorro. The outfit he wore was so neat and trim. He was a lawman, and he had a smoky kind of quality to his voice. When you heard his voice, you knew he was a good, solid character. You know he played the bad guy in some of Dad's earlier pictures. I liked Clayton a lot; he personified the Lone Ranger.

Ron: When did you first start taking collecting seriously?

Dusty: You cannot really get serious unless you want to sink a lot of money into it, and I mean a lot of money. You know we get calls from Roy and Dale collectors here asking for the value on a collectible. One guy said, "Well, how much do you think this item is worth?" "Well," I said, "it depends if it's the last one this guy needs in his collection, or if he already has six of the same thing." And that is the up and down of it in the collection business.

I do not collect anything that I just do not personally like. There are some of Dad's gun belts that I would give my eyeteeth to have today; but there are also some that I could care less about. And, you know it does not have to be the best one.

Now, if you are one of those collectors who collects only new in the box stuff, "Whoa," now you are talking about someone who needs an investment banker as a best friend.

Ron: What are your favorite things to collect?

Dusty: Probably pictures of my grand kids. (Laughs.) I have so many different things that I collect. I have fishing lures, old tools, and I like old tackle boxes. I used to go to auctions to buy old tackle boxes just to see what was in them. But, I am like Dad; if you look in the museum, you can see that there is not one specific thing that he really collected. He liked all kinds of different things. And, that probably came from an age when he did not have anything. So, when he got something he really liked, he just hung on to it and enjoyed physically having it. And, if you look around here, (Dusty's office in the museum) there is nothing specific, just a lot of stuff, just me and my stuff (laughs). When I go somewhere like a shooting meet, I will save the badge from that, or a back stage pass from a concert somewhere. Some day I will expand my exhibit window space here in the museum. Right now, there is not enough room in it, and I know my wife will force the issue. Right now a lot of it is here in the office. I have a gun collection, coin collection, and a stamp collection that I have pretty well given away.

I think that as we get older our priorities change, but I do not think that the little boy or little girl in us ever goes away, and what was important to us in our younger days is still somewhat important. We all try to go back to a time when we felt safe and secure and really enjoyed life. Back then, everything was just a pie in the sky. That's why the toys come to mind in all of us. You know, someone will say, "Yeah, when I was a kid I had that bike, or whatever, that brings back the happy memories." That is why the baby boomer generation with more disposable income is able to purchase expensive collectibles. They buy the best that they can get.

Ron: If you could only keep three things in your collection of western hero items, what would they be?

Dusty: One would be my first pair of Roy Rogers cowboy boots. I still have the original pair that Dad had made special for me. The next piece I would keep is a pair of Roy Rogers 24-karat gold-plated guns in their original gold gun belt. I have a Roy Rogers horseshoe money clip and I have never seen another one like it.

Ron: I heard that the Fanner 50 was one of your favorite guns when you were a kid?

Dusty: Yes, that is true. You know when the Fanner 50 came out, Mattel did a big television push on it. Well, I had to have one of those. The Schmitt cap guns were okay, but they did not hold a candle to the Fanner 50's. You could quickly blow off all the caps, and it would smoke really good. I still love the smell of caps. I still have the pair of Fanner 50's that I had as a kid. That is something else I would have a hard time giving away.

Ron: What is your favorite cast iron cap gun?

Dusty: The Roy Rogers Long Tom. They are really hard to come by, and I only have one.

Ron: What is your favorite die-cast cap gun?

Dusty: That would probably be a gold set of Roy Rogers Classy cap guns.

Ron: Dusty, what would you say are some of the most sought after Roy Rogers and Dale Evans collectibles?

Dusty: Probably for the guys, the guns and the holster rigs would seem to be the things to get. But, there are some items like the sterling silver rings that are very hard to find.

Ron: Is there any particular Roy Rogers collectible that personifies your father?

Dusty: I think the Classy gun rig that was done in the 1950s. I wore it all the time, and got pictures of myself and other kids wearing it. That to me, and the white hat that Dad wore, pretty much personified what Roy was all about for boys at the time.

Ron: Your Dad was a really big collector, who started collecting after his visit to the Will Rogers Museum. At that time, he had made a decision to hang on to things relating to his life and his career for his fans and put them into his own museum. Did Roy have favorite types of things to collect?

Dusty: I do not think there was anything specific. He collected mostly things that were given to him. He had a pretty good collection of sports memorabilia. He went to a lot of baseball games, especially the Yankees and the Cubs, where he would get baseballs signed by the players.

Probably what he collected which is not really socially accepted these days are the stuffed and mounted animals in the museum. He made two or three trips to Africa. Back then, it was a different time and there were different attitudes. He wanted people to come to the museum and see everything here. He had a good collection of guns. He had so many shotguns because he had a bent barrel on one of his 12 gauge guns and did not know it was the gun's fault, and not his, until he had bought about ten guns that he did not need. Dad was like a vacuum cleaner in collecting things. Whenever he went out, he just picked things up.

Ron: I read that in his later years your Dad loved to find "treasures" at flea markets. What types of things did he bring home?

Dusty: Oh, yeah, Mom called him a pack rat, and I guess he was. He had an affinity for pocketknives. When he was little, all he got for Christmas was a pocketknife and some fruit, an orange or whatever. The knife was not only a great gift, but also a workhorse, as Dad would go to the river and make whistles and sling shots. He was very good at killing small game with the slingshots. He would find ball bearings at the railroad tracks and use then to hunt rabbits and squirrels. That knife was very important to him. So, anyway, he loved to pick up knives. Another time he went through a watch-collecting craze.

Ron: Did Dale collect anything?

Dusty: No, she did not really get into collecting anything other than dolls. Now Mom has a charm bracelet and loves to get a new charm once in awhile.

Ron: Are your sisters collectors?

Dusty: My sisters used to collect little porcelain Disney characters. Walt Disney was a close family friend and would send all kinds of stuff. Boy, I wish I had kept that stuff, too! "Too soon old, too late smart." (Laugh)

My sister Linda inherited the salt and pepper shaker collection from my grandmother on our mother's side, and she has kept that collection going.

Ron: Is it true that you, Sandy and your sisters, when you were kids, "field-tested" the many Roy Rogers products that your family was given for your parents' "seal of approval"?

Dusty: Yes, Dad had that "Pledge to Parents" hand tag put on everything that was approved by him. It said that it was the best quality product and that the consumer was not paying any more for it just because his name was attached to it. So, a lot of time, companies would send this stuff to our home like shirts and pants, etc. We would wear that stuff all the time, and play hard in it like climbing rocks. So, we field-tested them. I remember my mom telling a jeans company, "You need to reinforce the crotch and the knees, because that's what my boys tore up the most."

Companies would send boxes of stuff. I remember companies that would send us gun and holster sets. Dad always wore the double rig, which the companies would send, but sometimes, they would send the Roy Rogers single rig. You know, not every parent could afford to buy their kid the double rig, so the single rig was all they could get their kid. Classy Products would send two or three dozen cap gun/holster sets out to the house; so, my Dad would keep a stack of them in his closet. Well, when we were playing cowboys, if we lost a cap gun, heck, we would just go in Dad's closet and get another one. (Laugh) Why, there are probably a thousand Roy Rogers cap guns all over the hills in Chatworth, where we dropped them between rocks and couldn't reach them.

We really were hard on stuff. Because we had so much of it, we would play with it constantly, especially the toys, clothes, etc. We tried everything, including food products. Everything kids had in those days, manufacturing companies would run by us. I guess they figured if the Rogers' kids could not destroy it, it must be pretty good. The bad part was that these companies would come out to photograph us. We could hardly do anything without a camera in our face. That was kind of a drag.

Ron: So, every day must have been kind of like Christmas?

Dusty: It was, but of course, we had to work for it. While they took pictures, we had to be good and cooperative, but when they left, we would tear around again.

Ron: Were there about 450 authorized items on the market at that time?

Dusty: Well, I heard different numbers, but there were ten unauthorized for every one that was. That is why now I see so many things that I have never seen before. Those companies did not come to us. Someone would make a copy of an authorized item and flood the market. By the time they were tracked down, they would be gone. So, who knows how many items there are legally and illegally licensed.

Ron: Do you remember any toys or other items that did not get Roy's approval for marketing?

Dusty: There was a ride on a Trigger horse with wheels. Well, all you had to do was sit on it and it broke. There was also an inflatable "Trigger" that you blew up and bounced on. We popped more of those things.

Ron: What kind of products would your Dad not lend his name to?

Dusty: Not too many things. I often wondered why he did not have something in the line of Red Ryder BB guns by Daisy. But, I guess Dad did not want kids shooting each other and putting an eye out. He did not want his name attached to anything like that. Also, he did not want his name on anything that was poorly made. He would not endorse anything that was flat out dangerous. I remember there was a TV chair that collapsed a lot. It had three legs and the thing kept breaking if a kid put a lot of weight on it. So, Dad told the manufacturer to fix it. People did not worry about safety as much then. Dad was also concerned about a toy box that would hit you on the head or catch your hand. He also would not put his name to something that just would not last.

Ron: How many new Roy and Dale products are currently being manufactured?

Dusty: We probably have about 40 licenses out there. They are not all toys. Recently we did a license with Daisy Manufacturing for a Roy Rogers BB gun. That's been a real popular item. Then there is Breyer Company that makes the Trigger and Buttermilk plastic horses. There's still a lot of interest, not so much in toys; but in more adult-oriented items.

Ron: Have you and Roy Rogers Enterprises had problems with companies that have recently produced unlicensed items?

Dusty: Yes, there have been people that would take advantage of a situation. Like someone has taken a regular watch and clock, and glued Roy's picture on them. With the laser technique it is getting easier to fake things. Like this guy selling boxes that he has put together and is getting money for it. They look pretty good, but he could have done it legally. When this happens, I do have my attorney send something telling them to cease and desist, but they are very hard to track down.

Ron: What are the best and the worst things about being Roy Rogers' and Dale Evans' son?

Dusty: There are not any worse things, really, but people do hold me to Dad's standards and that can be pretty hard to live up to. But that is fine, it gives me a goal to shoot for. That's probably the most difficult part, trying to be your own person. Mom said, "If you don't become your own man and deal with who your Dad is, you'll end up in a mental institution." My job is not so much to stand in Dad's shadow, but to lengthen it. So, that is how I see it.

The nicest thing is hearing the stories about Mom and Dad. To have had such a positive influence on thousands of children worldwide is phenomenal. Someone asked my Dad, "Roy, when did you first feel that you had a responsibility toward young people?" My Dad responded that when he was performing at a rodeo, after his act, this little kid came around the corner, and he was dressed up just like him. People felt that they could feel close to Roy. They respected him for the ideals he stood for. You know the first thing he would do when he went to a new town to perform? He would head for the local hospital and entertain the folks there. Once he was at a hospital for blind children. They could not see who he was, so Dad got down on his hands and knees and let the kids touch him, his clothes, hat, guns, so that in their mind's eye they could see who he was. Now, how many stars today would do something like that? Dad just loved doing things for children. It was really hard for him to see kids racked with polio and other diseases. I know when my sister Debbie was killed, Dad was angry with the Lord for taking his child. It was so hard for him to understand, and he had a difficult time dealing with it. He also could never understand how parents could neglect their own children. So, anyway, he wanted to make a difference in children's lives, for the better.

Ron: I know that your Dad loved to kid people. Do you remember any funny stories about Roy playing a joke on someone?

Dusty: Dad had a great sense of humor. He loved to tell stories and pull stuff on people if he could get away with it. That would really tickle him. But the most tickled I ever saw him was when someone would pull a trick on him. With the kind of life he led, he needed to be able to laugh once in a while. Hollywood was tough and did not fit him well. I remember one time he pulled a stunt on his film crew. They were all out filming up in the rocks in the valley on Iverson Ranch. Well, during the break Dad had gone with a buddy and they had found a deep hole by some rocks. Dad told his buddy that he was going to holler out like he had fallen in, then you go back and tell the crew that I've fallen into this deep hole and I'll hide behind the rocks. So, the whole crew came carrying ropes and ladders and went out to rescue Dad. While they were setting this stuff up and yelling down the hole to see if he was okay, Dad was behind a rock, laughing like crazy. He would get such a kick out of pulling stunts like this. He would not do anything to hurt ya, but he would put salt in your coffee when you were not looking just to see your face when you drank it.

Ron: Roy's legacy has been so enduring and spiritually uplifting because his love was so genuine. He seemed to have brought out the best in all of us. Do you sense that his spirit is still watching over you and your family?

Dusty: Absolutely. This last father's day, I had some trouble with this problem I had to work out. Since I always used to bounce things off of Dad, I thought that I would go up to where he was buried and ask for his help. I said, "Dad, I'm having this trouble with an idea. If you have any suggestions, let me know." When I left his grave site and got a couple of miles away, I had this feeling that he was with me and telling me just to remember what we've done in the past. Remember and go with that. You just know what's right, now go and do it. That really helped me a lot.

Then there was a time when we had a photographer upstairs in the museum and we were shooting Dad's things that he really liked for a poster. We had his guns, hats, watches, ring, belt, etc., all laid out for the shooting. So, after this was done, everyone was gone, and I was closing up. I checked doors and locks first and then saw Dad's silver belt, and I thought if Dad was here, he would have me put that in the safe where it should have been. He did not like to have stuff like this lying around. But, I thought, "Oh, I'll do it later." So, I went ahead and punched in the security code, and it wouldn't set. So, I went back upstairs, went to the door with a flashlight and the light went across the silver belt. Okay, I understand, I grabbed the belt and put it in the safe, then punched in the code and the security system set fine. So, now he's around, and I think the good Lord makes it that way. Yes, of course, we miss him, but there's a part of him that stays around to help fill the void so we don't feel completely alone. If this feeling comes from my heart or my head, I don't know; but it doesn't really matter. You know Dad was a practical joker and he does that here. The ladies will set up a display in the shop and the next day the display will be back to where it was originally. So, they'll set it back up again, and the next day some things will have been moved. Yes, Dad likes to mess with people like that. I can just hear him giggling about it.

Ron: What advice do you have for new collectors?

Dusty: I would say collect what you really enjoy. Find something specific you like and go for that, unless you are doing it strictly as an investment. Right now Roy Rogers items are hot and baby boomers are buying them up. With the Internet, exposure to more people makes things more valuable. Any price guide is going to be difficult as prices change so quickly. About all one can do is give guidelines.

ROY ROGERS AND DALE EVANS MUSEUM

Roy Rogers' long time dream came true in 1967, when the first Roy Rogers Museum was completed and opened to the public in Apple Valley, California. Roy had made a decision many years before to collect and save as many mementos from his professional and personal life as possible for his own museum. He had decided this after visiting the small and sparsely furnished Will Rogers Museum.

Roy Rogers Museum in Apple Valley, California. 1967–1976.

A new and larger Roy Rogers and Dale Evans Museum with 32,625 square feet of floor space was completed in 1976. This huge fort-looking museum is located just off Interstate 15 in Victorville, California. The Fort Apache-looking museum situated in the high desert landscape takes one right back to the days of the Wild West.

Roy Rogers and Dale Evans Museum.

Entrance to the Roy Rogers Museum.

A rearing Trigger, standing two stories high, greets visitors at the museum entrance.

Once inside the museum, you will be amazed to see the wonderful artifacts and memorabilia that tell the story of Roy, Dale, their family, and their film and television co-stars. Their professional and personal history dating back to their roots comes alive in viewing the many exhibits, displays, and works of art. There is even a theater to watch a Roy Rogers and Dale Evans movie. One of the museum's main attractions, Roy's horse Trigger, is featured in an outdoor scene exhibit with Buttermilk and Bullet. You can feast your eyes on beautiful parade saddles, including the really fancy one that is laden with silver and sparkling jewels that cost Roy over fifty thousand dollars back in the 1950s. There are displays of colorful costumes that Roy and Dale wore, awards they received, family photos, fine art, and numerous other exhibits and displays. You can even see the old truck that brought the Slye family out to California, and Pat Brady's jeep "Nellybelle."

Pat Brady's jeep "Nellybelle" and little "Nellybelle" pedal car.

A small part of Roy and Dale's beautiful western costumes exhibit.

Roy Rogers' western customized Pontiac Bonneville convertible with inlaid silver dollar interior and six-shooter door handles.

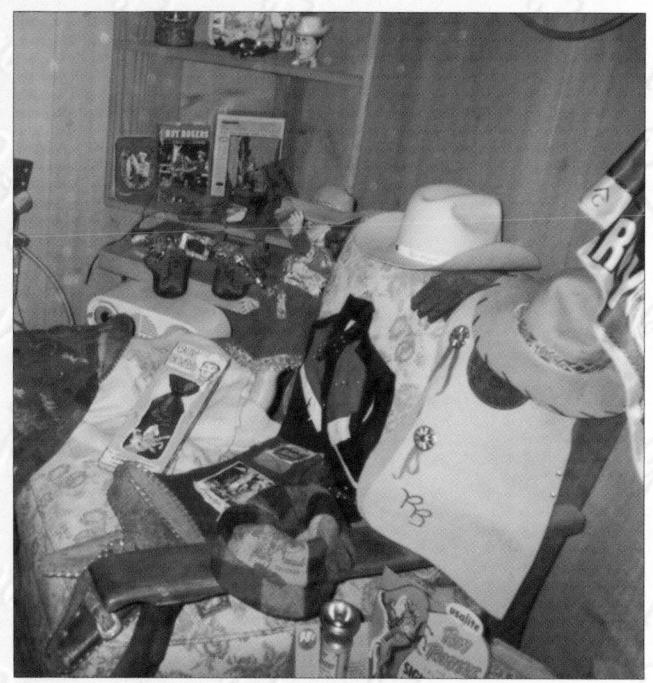

A small part of the Roy Rogers children's clothes and toys display.

Unlike most western museums, the Roy Rogers and Dale Evans Museum has a special warm and homey charm that makes visitors feel right at home with the Rogers family. Roy used to come by the museum most mornings to greet visitors who were absolutely thrilled to meet and have their pictures taken with him. Dale would do the same thing. A few years back, my wife gave me a trip to the Roy Rogers and Dale Evans Museum for my birthday present and I, of course, was really excited that I might meet the "King of the Cowboys." We arrived at the museum as soon as it opened and discovered that Roy was in the hospital and wouldn't be visiting the museum for awhile. I was disappointed after looking forward so much to meeting him, but was soon totally absorbed in looking at all the great displays, saying, "Wow! Look at that" every five seconds. I didn't even realize my wife, Roxanne, was in another part of the museum and I was talking to myself. About ten minutes later, Roxanne came running up to me and said, "Ron, come quick!" So I followed her back to the museum lobby and sitting there surrounded by admirers, was Dale Evans, "The Queen of the West." My heart thumped so loud at seeing her that I thought that everybody probably heard it. Wow! She looked beautiful with her delicate features and sparkling blue eyes, and radiated genuine warmth and kindness. I am so very fortunate to have met such a wonderful lady and remember those moments well.

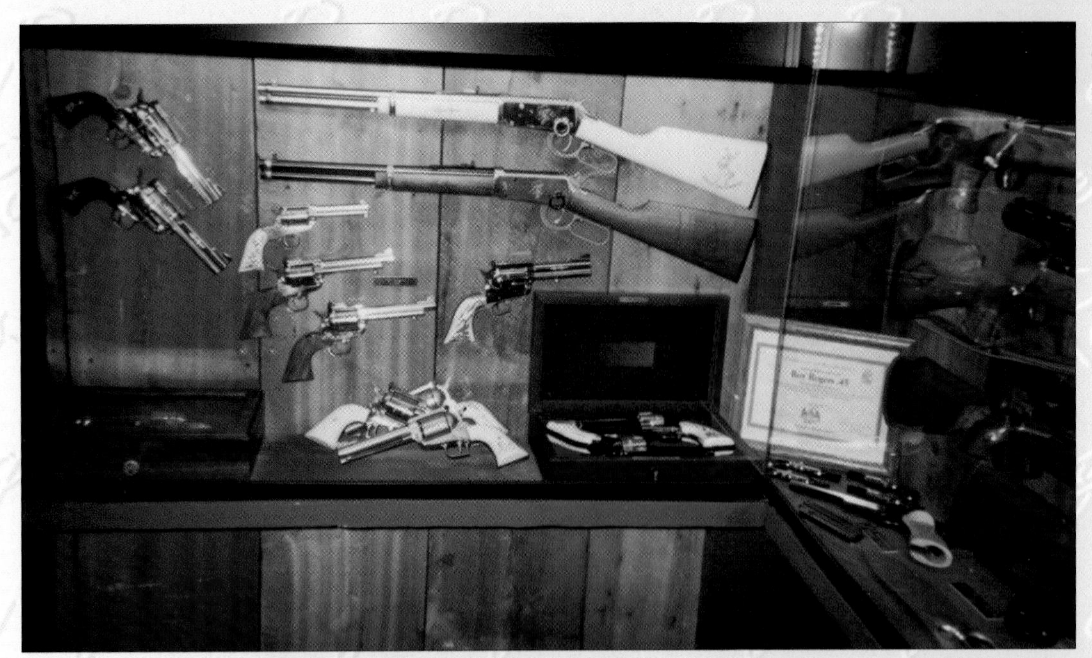
Some of the guns in the impressive Roy Rogers gun collection.

A small sample of the numerous awards received by Roy and Dale.

I encourage everyone, young and old, to visit the Roy Rogers and Dale Evans Museum as it is truly a wonderful learning and entertaining experience. There is so much to see and enjoy that you will want to come back many times.

Roy Rogers' son, "Dusty," and the staff of the museum are super-nice people and are very helpful in answering visitors' questions.

For more information about the Roy Rogers and Dale Evans Museum, you can log on to www.royrogers.com on your computer or call (760) 243-4547. The museum is open every day except Easter Sunday, Thanksgiving Day, and Christmas Day from 9:00 AM to 5:00 PM. There are five nice motels and inns located within a two mile radius of the museum. If you have a self-contained motorhome or travel trailer, you can stay on the museum property overnight for free.

ROY ROGERS "KING OF THE COWBOYS"

The western hero that we all know as Roy Rogers was named Leonard Frank Slye by his parents, Andrew and Mattie Slye, when he was born on November 5, 1911, in Cincinnati, Ohio. At the time of his birth, Leonard had two sisters, Mary Elizabeth and Cledda May. A baby sister, Kathleen, born in 1915 completed the family.

For the first seven years of his life, Leonard and his family lived on a houseboat built by his father and Uncle Will. After saving up enough money to purchase some land, the Slye family traveled up the Ohio River and bought a small farm of several acres near Duck Creek, Ohio.

During his early youth, Leonard helped his dad, Andy, and uncle build a six-room clapboard farmhouse that had no indoor plumbing or electricity in this hilly brush country. After trying for awhile to eke out a living from the soil without success, Andy returned to making wages as a shoemaker at a factory in Portsmouth, Ohio. The rest of the family was put in charge of running the farm as Andy could only be home on the weekends.

Leonard learned about the hard work that went into farming; but had also gained some valuable experiences that shaped his character and helped him later in life. His first horse riding experience was on a black mare named "Babe" that was given to him by his dad. Leonard became an accomplished rider and rode Babe to school, prayer meetings, square dances, and on occasion to the Portsmouth Movie Theater to see his favorite cowboy hero Hoot Gibson.

At the tender age of twelve, Leonard won a horse race with Babe at the Ohio County Fair at Scioto. He also got first prize in the 4-H contest for raising a black Poland China pig that he had named Martha Washington. The prize was a one-week trip to the 4-H Club Congress in Columbus, Ohio, where he experienced city life for the first time.

The Slyes were a very close-knit family, and their home was filled with love and music created by all the family members. Andy and Mattie taught their children how to sing, yodel, play guitar, and call square dances. All were valuable lessons for the farm boy who was destined to become a Hollywood star.

During the height of the Great Depression in 1930, the Slye family used their meager savings and headed for the "Promised Land" of California. This trip with its problems of car repairs to their 1923 Dodge was a scene right out of John Steinbeck's, *The Grapes of Wrath*. In California the family stayed with Leonard's sister, Mary, who had previously moved there with her husband.

Jobs were scarce then, so after four months, the family traveled back to Ohio. In 1931, the Slye family sold their Ohio home to some neighbors and moved lock, stock, and barrel back to California. Leonard and his dad got jobs driving Model T dump trucks and picking peaches in the San Joaquin Valley. Andy and Leonard would entertain the workers with their guitar playing and singing after a hard day working in the peach orchards.

Leonard's goal was to make a living playing music. He played guitar and sang on local radio shows and stage performances for free just so some talent scout would notice him. During this time, "Len" as his friends called him, formed several cowboy-type bands, including one called "The Sons of the Pioneers." This group got their big break when they signed a contract with Decca Records in 1934. One of their top recording hits was "Tumbling Tumbleweeds" which reached number 13 on the charts.

Leonard's first marriage to Lucille Ascolese ended in divorce after only three years. He felt that his being on the road with the band probably contributed to its failure. His second marriage was to Arline Wilkins in June 1936. Leonard's claim to fame happened when he had heard about a screen test at Republic Studios for singing cowboys while getting his only hat cleaned in a hat store in Glendale. He rushed to the studio and convinced producer Sol C. Siegel to give him an audition. Mr. Siegel agreed to the audition because he knew Len was one of "The Sons of the Pioneers." It turned out to be a successful screen test for Leonard and he landed a contract with Republic. Gene Autry was the big western star at Republic during this time. Republic Studios gave Len the new name of Dick Weston and used him as leverage to get Gene Autry to sign a new contract. Shortly thereafter they changed Len's name to Roy Rogers, a name Len liked because he was a great admirer of Will Rogers.

Under Western Stars was Roy's first feature film produced by Republic Studios at Lone Pine in 1938. Two years later Roy hired Art Rush to serve as his personal manager and agent. Art played a major role in Roy's success throughout his career and they became the best of friends. At this time Roy made eight films a year and traveled over one hundred thousand miles a year touring the United States entertaining folks. With Gene Autry in the army, Roy became The #1 Singing Cowboy in the movies. He was also bestowed the title, "King of the Cowboys," and voted #1 Western Star by Hollywood film magazines. He remained in this position until the magazine polls ended in 1954. His movies were shown at over seventy-five hundred theaters across the country and Roy traveled over fifty thousand miles a year entertaining at rodeos, hospitals, and army bases. Roy Rogers comic books in a new four-color medium came out in 1944, the same year Art Rush created a huge merchandising program that included clothes and toys with the Roy Rogers name. Roy's popularity soared. Whitman Books produced Roy Rogers books and music companies produced *Roy Rogers Song Books*. By 1946, 700 Roy Rogers fan clubs formed in the United States and 90,000 fans were writing to him monthly. His popularity also extended to other parts of the world where his picture was placed on the cover of leading foreign magazines.

Roy's family life was expanding as well as his career. The Rogers family had already adopted a little girl, Cheryl Darlene. Several years later when Roy was 30 years old, Arline gave birth to their new daughter, Linda Lou Rogers. Roy was so happy that he wrote a song just for her.

In 1946 Roy Rogers, Jr., nicknamed "Dusty," became the third child of the Rogers family. Six days later Arline passed away, dying from an embolism. The pain of this unexpected and shocking event was one of many that Roy suffered in his life. Besides losing great friends like Art Rush, Pat Brady, and Gabby Hayes, he also endured the pain of losing three of his children, Robin, Debbie, and Sandy.

Dale Evans was under contract with Republic Studios and had met Roy Rogers at Edwards Air Force Base in Lancaster, California, where they were both performing for the troops. She became Roy's leading lady in Republic's movie *The Cowboy and the Señorita* filmed in 1944. The "King of the Cowboys," Roy Rogers, and "The Queen of the West," Dale Evans, were united in marriage on December 31, 1947. This was truly a marriage made in heaven that survived through thick and thin when most Hollywood celebrity marriages failed. They had opened their hearts and their home by adopting six of their nine children. Their children are Tom, Roy Rogers Jr., Cheryl, Marion, Sandy, Linda, Dodie, Debbie, and Robin.

Roy and Dale shared their movie and television spotlight with co-stars Trigger, the "Smartest Horse in the Movies" (with Roy since 1938); Buttermilk, Dale's horse; Bullet "Roy's Wonder Dog"; and sidekicks Pat Brady, Smiley Burnette, and Gabby Hayes. They rode into our homes via the television in *The Roy Rogers Show*. Over one hundred episodes were televised between 1951 and 1955. The radio version of *The Roy Rogers Show* was produced from 1944 to 1955. The rest is history as Roy Rogers went on to more fame and fortune with the Good Lord guiding his way. He and Dale became the models for parents and family values.

From his humble beginnings in Duck Run, Ohio, to his star status in Hollywood, California, Roy never forgot his roots and remained the same spiritual and loving person both on and off the screen. His Christian beliefs and love of family supported him throughout his life. He freely gave his time, energy, and love to countless numbers of children and adults around the world. Roy became the world's most celebrated and loved western hero movie star, and lived every minute to the fullest. His sense of adventure propelled him to fame, starring in 88 musical westerns. He was the number one box office star in western movies for twelve years in a row, and recorded over 350 songs.

Roy Rogers' life was as full as any of us could ever hope to have. He was a man's man who loved to go hunting, fishing, camping and hiking. Roy also did some boxing, shot pool, bowled, rode his own motorcycle, was an avid trap shooter, did most of his own movie/television stunts, collected rocks, raised racing pigeons, and raced hydroplanes. Roy also loved animals, and always had a lot of hunting dogs and other critters on his ranch. For several years he even had his own rodeo.

Later in life, Roy loved to greet his fans at The Roy Rogers and Dale Evans Museum. Another of his special things to do was to go to flea markets and collect "special" stuff. He greatly loved his huge family and really enjoyed their time spent together. One of his grandkids asked him to yodel, and Roy did his last yodel three days before he passed away in his sleep.

To lady admirers Roy was a real "sweetie pie." He was a very handsome man with a twinkle in his blue eyes that made women swoon. He also really loved sweets, anything with peanut butter, caramel, or butterscotch. Some of his favorites were "See's Candies" and "Werther's" gold-foiled butterscotch candy. He never seemed to be without a stash of it in his pockets. As a symbol of their love for their dad, Roy's kids put a piece of candy in his pocket at his funeral "to take with him to heaven."

Roy Rogers passed away on July 6, 1998. His legacy will remain forever in our minds and in our hearts, as he represented honesty, decency, courage, love of country, people and the goodness in all of us. He led an extremely purposeful life and we miss and love him.

DALE EVANS "QUEEN OF THE WEST"

Dale Evans was born in Uvalde, Texas in 1912, and given the name Frances Octavia Smith by her parents Walter Hillman and Bettie Sue Woodsmith. As a little girl, she had dreams of one day getting married to Tom Mix. She attended business college after high school, and got a job as a claims adjuster. Frances loved to sing around the office. Her boss recognized her singing talent and helped her get started in her own radio program.

Frances had a son, Thomas Fredrick Fox, Jr. from her first marriage. This marriage ended in divorce when her husband deserted her and she received custody of her son. In 1934 she was occasionally performing with the stage name of Marion Lee. Her program director, Joe Eaton, suggested that she use the name "Dale Evans" and this became her professional name.

Dale worked very long, hard, hours with a lot of traveling as a singer to make a living. By 1940 she had remarried and was living in Chicago with her husband and her son Tom. Dale was singing at various nightclubs and had her own show *That Girl from Texas* on CBS Radio. A year later, she received a one-year contract with 20th Century Fox Studios. Republic Pictures gave her a long-term contract in 1942 where she made eleven non-western films.

Dale Evans' first western movie was with John Wayne entitled *War of the Wildcats*, also known as *In Old Oklahoma*. She appeared with Roy Rogers in their first movie together entitled, *The Cowboy and the Señorita*, which was filmed in 1944. Dale continued to be the leading lady in other Roy Rogers movies. She also made many personal appearances with Roy and later became a friend of the Rogers family.

Dale filed for and received a divorce from her husband, Robert, in Los Angeles in 1945. Her friendship with Roy grew into a love relationship and they were married on the last day of 1947. Soon after, Dale made a conscious choice to live the life of a Christian. Roy soon followed the same path.

The Rogers family experienced great happiness and much heartbreak in their lives. Dale has written extensively of these events in her books. She has dedicated her life to her faith and family, while maintaining her career as the one and only Dale Evans, "Queen of the West."

Dale Evans passed away at her home in Victorville, California, on February 7, 2001 at the age of 88. This very special lady is greatly missed.

GEORGE "GABBY" HAYES

Gabby Hayes began his acting career in his youth, often portraying an old man in a vaudeville theatre. Hence, he got a reputation of being born old. He has also been called "the old dean of western movies" playing the role of sidekick to several famous screen and television cowboy stars. Gabby teamed up with Roy Rogers as his sidekick in their first movie together, *Southward Ho* in 1939. Roy and Gabby made 41 films and numerous television shows together. In 1950, Gabby had his own television show on NBC, *The Gabby Hayes Show*. Gabby also was on the radio show *Weekly Roundup* in 1946 with Roy, Dale, and The Sons of the Pioneers.

Mr. Hayes was a true actor as he assumed a completely different persona when he acted the role of a grumpy old geezer in a western show. In this role he would say things like, "Dag nab it!" and "Young whipper snapper!" In real life, Gabby was a man of great intellect, style, and class. George would arrive at Republic Pictures Studios in Hollywood for the day's shooting driving a new Lincoln and wearing a fine suit complete with derby hat. After a change of clothes and hat, he would emerge from his dressing room as that cranky old cuss, Gabby.

Gabby was a great acting teacher and gave Roy and other cowboy stars a lot of sound advice in their profession. By working together so much, he and Roy developed a very close and lasting friendship. Roy said that, "Gabby was like a father, brother, and buddy all rolled into one." He was a master of spinning yarns of courage and adventure, and did a lot to keep the traditions of the Old West alive and well. George "Gabby" Hayes died in 1969, leaving us great memories of his acting career in western movies, television, and radio shows.

PAT BRADY

Pat Brady, "the man with a rubber face," as quoted by Roy Rogers because of Pat's facial expressions, was born on December 31, 1914. His Irish parents, Jack and Tete Brady, were actors in Broadway shows in New York. As a young man, Pat left New York, became a musician, and ended up playing bass fiddle in Roy Rogers' musical group, "The Sons of the Pioneers." This group appeared in many of Roy's earlier movies. Pat played the role of Roy's sidekick in five movies starting in 1949 with *Golden Stallion* and ending in 1951 with *South of Caliente*. He was also the sidekick in over one-hundred Roy Rogers television shows in the 1950s. In the TV shows, Pat drove an old military jeep named "Nellybelle," and played the role of "Sparrow Biffle." Sparrow was a cook in Dale Evans' Eureka Café on the television set of Mineral City.

Pat and his wife, Fayette, were very good friends of the Rogers family and served as godparents for Roy and Dale's daughter, Dodie. In February 1971, Pat had a heart attack and passed away, but the humor, musical, and acting talent of this red-haired, freckled-face man lives on in our memory.

NOTE: Other sidekicks of Roy Rogers were Gabby Hayes, Andy Devine, Smiley Burnette, Gordon James, and Pinky Lee.

TRIGGER "THE SMARTEST HORSE IN THE MOVIES"

Roy Rogers' intelligent and beautiful palomino horse, Trigger, was born in Santa Cietro, California, in 1932. His owner at that time, Mr. Roy Cloud, had given him the registered name of "Golden Cloud." His dam was a cold-blooded palomino and his sire was a thoroughbred racehorse. Golden Cloud began his life as a racehorse. He was ridden by Olivia DeHavilland in the 1938 movie *The Adventures of Robin Hood*, starring Errol Flynn. Trigger was part of a stable of horses that were used in the filmmaking business. Republic studios rented Golden Cloud from Hudkins Stables and used him in Roy's first movies. Roy's sidekick in these early films, Smiley Burnette, came up with the name "Trigger," for Golden Cloud. Roy liked the name and loved the horse so much that he purchased him from Hudkins Stables for $2,500 and used Trigger in his films and television shows.

Jimmy Griffin was the first trainer of Trigger hired by Roy. Glen Randall was the trainer that taught Trigger over 60 tricks. Glen trained Trigger and several Trigger doubles from 1941 to 1965. Trigger was a real ham, and Roy said, "Trigger could spin on a dime and give you nine cents change." Trigger made many personal appearances with Roy, including the New York World's Fair. One of Trigger's publicity stunts was to walk into the hotel where Roy was staying, and sign his name with an "X" at the registration desk. Because of his affection for Trigger, Roy signed Trigger's name with his own when giving out his autograph.

At one time the rumor had gotten out in the newspapers that Roy was going to sell Trigger to some big tycoon with lots of money. Well, Trigger's young admirers heard about it and sent in quarters, and whatever they could give to Roy so he would not sell Trigger. Roy was so touched by this that he wrote them all a letter telling them that Trigger would not be sold. He also sent thousands of honorary Trigger Ownership Certificates to the children who mailed in a request.

Trigger, Jr. and Trigger doubles were used in many personal appearances, movie stunts, and some television scenes. The original Trigger was known as the "old man" and whenever a stand-in "Trigger" double could not perform a difficult stunt, the "old man" would get the job done. "The Smartest Horse in the Movies" proved he was a real star in movies where he got top billing in *My Pal Trigger* and *Golden Stallion*. He starred in all the Roy Rogers movies and in all the Roy Rogers television series. Some Trigger collectibles include Trigger comic books #1 through #17 that were published from 1951 to 1955, a pull toy made in the late 1940s, several wood rocking horses and an inflatable toy horse that cost $1.98 in 1955. All the Rogers family rode Trigger, except Robin, at one time or another because he was such a good-natured and gentle horse.

Trigger died in 1965 from natural causes, at the age of 33 at a ranch in Thousand Oaks, California. When Trigger passed away, Roy said it was like losing one of his own kids, and did not tell anyone about it until a year later. Roy made the decision to have Trigger stuffed and mounted so that people could see what an absolutely beautiful horse he was. Trigger continues to be one of the most popular displays at The Roy Rogers and Dale Evans Museum in Victorville, California.

BULLET "ROY'S WONDER DOG"

Roy Rogers loved animals and always kept a lot of them on his ranch. He raised palomino ponies, and in 1948 owned eleven hound dogs used for hunting and two very special dogs. One was a white German shepherd named "Spur," who appeared with Roy on many Dell comic book covers. The other was a German shepherd, who would have a starring role with Roy, called "Bullet." Bullet was featured in several of Roy's early 1950s movies, and in the Roy Rogers television series. His image was also shown with Roy on comic book covers and on coloring books.

As with Trigger, Bullet had doubles that filled in for him in action stunts that called for an attack dog to subdue the bad guys. Bullet collectibles are quite scarce because there were not many made. In the late 1950s, Hartland Plastic, Inc. produced a beautiful Bullet figure that is one of my favorites.

CONDITION EVALUATION AND PRICE GUIDE

As with most collectibles in today's market, it has become very difficult to keep up with market prices on Roy Rogers and Dale Evans collectibles and memorabilia.

Supply and demand are always the governing factors that determine the price paid for collectible items. These prices usually will vary from one part of the country to another and from one week to another. With this rapidly expanding electronic and computer age, more people are being exposed to collectibles through auctions such as eBay and television shows like *Antiques Roadshow*. Through this type of media, more people become collectors, and present collectors are able to enlarge their collections by bidding on auction items in the comfort of their homes.

Prices for Roy Rogers collectibles have been escalating for the last couple of years. The biggest rise in prices is for mint-in-the-box items and "new" old stock collectibles that have been discovered in some old warehouse, or in a closed department store basement.

Collecting Roy Rogers and Dale Evans vintage items is a fun hobby, and a great investment with the current price increases. Many current collectors consider their collections as a retirement account, and a more enjoyable one than stocks and bonds. Having a collection brings out the child in all of us, and instantly takes us back to our childhood when we felt safe and secure in a less complicated world. Just shooting a cap off in a cap gun and smelling the whiff of gunpowder brings back the memories of playing cowboys as a kid.

Back in the good ol' days of the 1950s Roy Rogers items cost more than many generic items and many of our parents just could not afford to buy the more expensive item. As an example, one type of Roy Rogers double holster and cap gun set sold for $4.95, whereas a comparable generic set sold for $2.99. As adults, we can now fill that void from our childhood and purchase that Roy Rogers collectible without asking permission. Well, some of us can anyway.

Some of the most important aspects of assembling a Roy Rogers collection are:

1. Collect what you enjoy the most.
2. Purchase the item in the best condition that you can afford.
3. Search for the best price for that particular item.
4. Keep upgrading your collection as your budget allows.
5. Offer to trade, i.e. trade up, things in your present collection that no longer appeal to you for things that do from other collectors.

I would advise new collectors to seek out the long-time Roy Rogers collectors for their expertise and valued advice, as most of them are more than happy to share their experiences with someone getting started. It is also important to gather as much information as possible about the items you wish to purchase for your collection before writing that check, so you will be making a well-informed and intelligent decision.

Be aware that there are many unauthorized items and reproductions on the market, now more than ever. Knowledge and experience will be a big help in detecting fake items.

Condition and rarity are perhaps the two biggest factors in determining an item's value. Of these two factors, rarity is difficult to deal with for a collector. When you see an item that you have not seen in any Roy Rogers collectible books, or in friends' collections, you have to do some research on it and find out if it is the genuine article. Ask a lot of questions and closely scrutinize the item before purchasing. You do not want to get stuck with an item that was made last year, but looks like it came right out of the 1950s.

The second factor for placing a value on a collectible is its condition. As Roy "Dusty" Rogers,

Jr., told me, "If you plan on collecting only mint-in-the-box items, you'd better have a banker as your closest friend." From my experience, this is very true, as the boxes sometimes cost as much or more than the item.

When considering condition, restoration may also come into play in the evaluation of an item. I have seen some restored Roy Rogers collectibles that looked better in their restored condition than when they were new, and other restorations done by amateurs who ruined the item by working on it. As a purist, I do not feel that finely restored items will ever be worth as much as originals in mint condition. However, I am really thankful that some very talented people are able to restore and preserve items that otherwise may have gone unnoticed and deteriorated. All sellers of restored items should disclose any alterations of the item to potential buyers.

Sometimes the history of an item will also add to the value of a Roy Rogers collectible. I know many collectors that would pay more for an item that came from Roy or Dale personally.

The price range values for items listed in this book are to be used only as a guide. These values were obtained through researching the latest Roy Rogers auctions, value guidebooks, and a panel of highly regarded dealers and collectors. There are simply too many variables to establish an absolutely firm value on any given item. Please use this book only as a guide when buying or selling Roy Rogers memorabilia and collectibles. The true value of an item is determined by how much you really want it, and are willing to pay for it. Or, as a seller, how much attachment you have for the item, and how much money are you willing to take for it.

In this book, I have given a price range value for each item. The lower price is for an item that is classified in a C-8 condition, which means that the item is in very good played with condition with no broken or missing parts, tears, or breaks. The high price is listed for any item that is in C-10, unused, never played with mint condition.

Please use your own good judgment when hitting the happy trail of collecting Roy Rogers and Dale Evans collectibles and memorabilia.

Happy hunting to you all!

ABOUT THE AUTHOR

Ron Lenius was born in Sturgeon Bay, Wisconsin, in 1943, and spent his youth living on various farms in Door County. He received a B.F.A. degree from California College of Arts and Crafts, and a M.F.A. degree from the University of California at Berkeley. For the past thirty years, Ron has been in the jewelry and gemstone professions. He has been collecting Roy Rogers and other western hero collectibles for many years. His other extensive collections include vintage scenic western guitars and Hubley airplanes. Ron and his wife, Roxanne, live in Sonoma, California.

Ron Lenius with Dale Evans.

CHAPTER 1

ADVERTISEMENTS

Roy Rogers Enterprises, founded by Roy Rogers, granted licenses (manufacturing rights) to companies that produced items that bore the name Roy Rogers. Advertising these products through various forms of media that reached the public was a vital part of selling a Roy Rogers item.

Large, full-color ads were placed in popular magazines such as *Life* and *Ladies' Home Journal*. Ads were also placed in major newspapers, catalogs, and other media. Products bearing the Roy Rogers name were also advertised nationally through retail store promotion programs.

Many of the Roy Rogers items carried a hang tag, "Roy Rogers Pledge To Parents," that stated the item was equal in quality and value similar to items in the same price range. This simple pledge instilled confidence in the consumer to purchase the product. Advertisements for Roy Rogers items were placed inside comic books and on the outside of cereal boxes. Roy Rogers became the "King of Merchandisers" with the aid of these various marketing tools.

1952 ad promotions for retailers. (R.L.)

Value $25-125

1950s Eastman Kodak Company ad. (R.L.)

Value $8-15

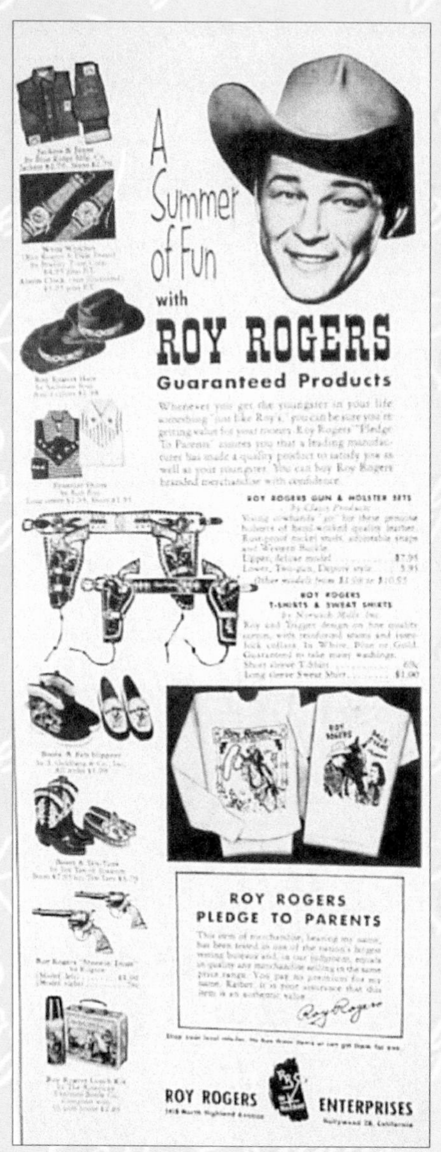

1954 Roy Rogers toy ad. (R.L.)

Value $5-10

1950s Dell comic book ad. (R.L.)

Value $3-5

1957 Roy Rogers full-page Christmas ad, 10" x 13 1/2", *Ladies' Home Journal*. (R.L.)

Value $12-18

1950s Roy Rogers Quaker Oats ad, full-page, from a Sunday newspaper. (R.L.)

Value $24-30

1957 Roy Rogers full-page Christmas ad, Sears, Roebuck & Company (R.L.)

Value $12-18

Original photo of Roy Rogers Cookies ad on semi-tractor/trailer. (Roy Rogers Museum)

Roy Rogers Cookies large full-color newspaper ad. (G.S.)

Value $60-85

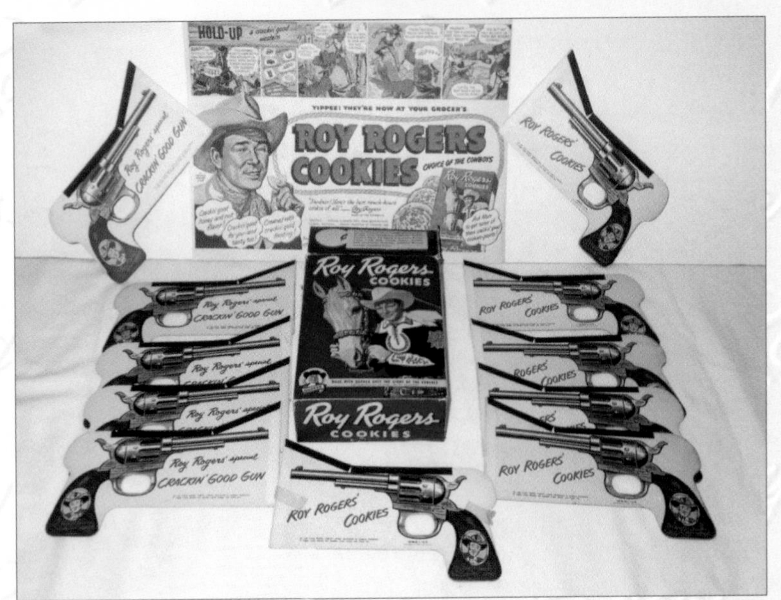

Roy Rogers Cookies ad with box of cookies and eleven eight-inch cardboard Crackin' Good Guns. (G.S.)

Value $45-85 each gun

Value $350-450 cookie box

Roy Rogers Cookies box with gun and premium game card. (M.M.)

Value $20-25, game card ad only

Roy Rogers "Family Contest" ad for Post Sugar Crisp cereal. (R.L.)

Value $4-7

Roy Rogers "Win Your Own Pony" ad for Post Toasties cereal. (R.L.)

Value $7-10

Roy Rogers Cookies box, very rare. (G.S.)

Value $350-450

Roy Rogers "Win a Palomino Pony" ad for Nestle's Quik. (G.S.)

Value $125-150

Roy Rogers "Trigger the Third" contest ad in Des Moines Sunday newspaper. (G.S.)

Value $125-150

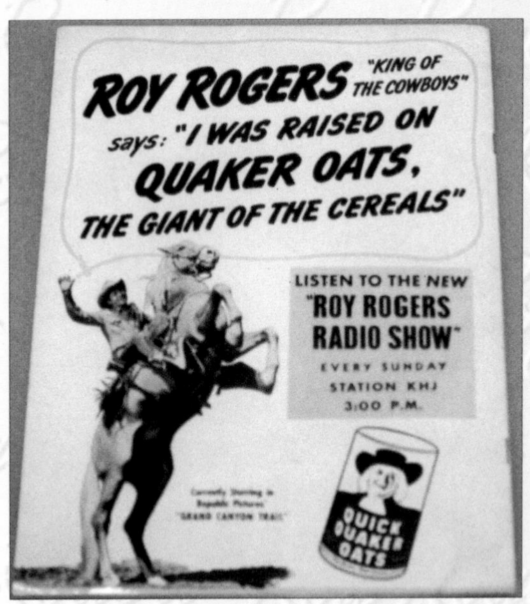

Quaker Oats cereal ad. (R.L.)

Value $3-5

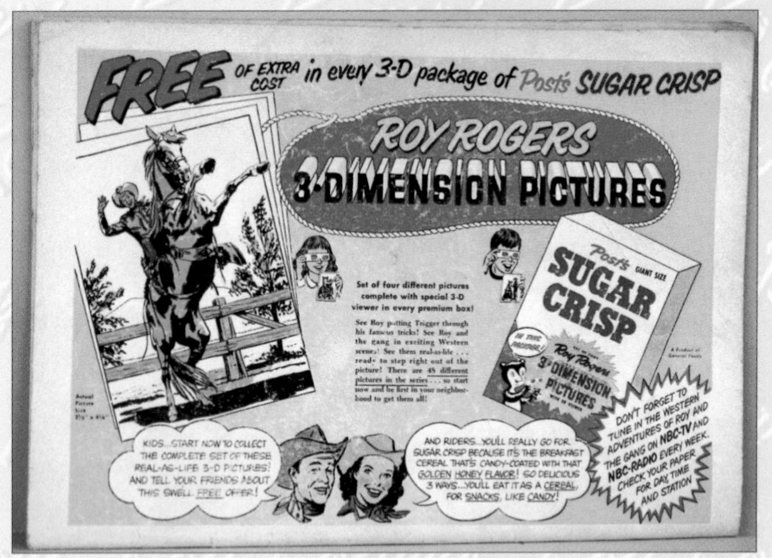

Post Cereal "Sugar Crisp" ad. (R.L.)

Value $4-7

Post Cereal "Grape-Nuts Flakes" ad. (R.L.)

Value $4-7

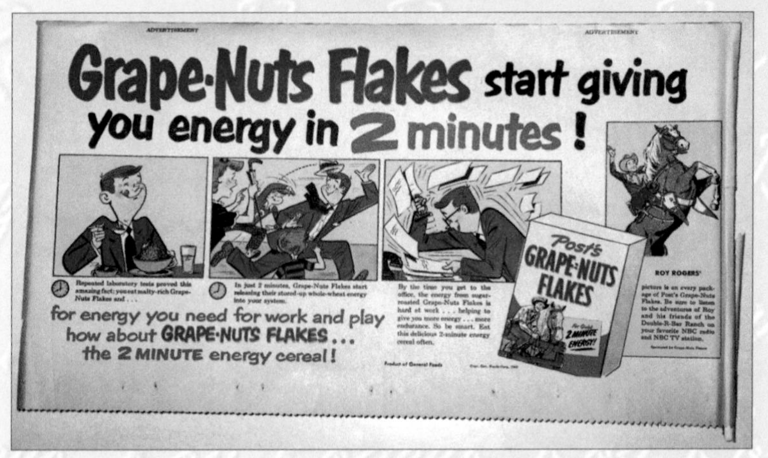

Post Cereal "Grape-Nuts Flakes" ad. (G.S.)

Value $4-7

Roy Rogers catalog ad for retailers. (Roy Rogers Museum)

Value $10-15

Premium offer ads for various cereal companies. (G.S.)

Value $4-7

Post Cereal "Grape-Nuts Flakes" ad. (R.L.)

Value $4-7

Quaker Oats premium offer ad. (G.S.)

Value $4-7

Post Cereal "Puffed Wheat Sparkies" ad. (G.S.)

Value $4-7

Roy Rogers two-page catalog ad for retailers. (Roy Rogers Museum)

Value $10-15

Page from Schwinn bicycle booklet. (Roy Rogers Museum)

Value $12-18

Classy Gun and Holster Catalog for retail merchants. (G.S.)

Value $90-150

Roy Rogers and Dale Evans retail store catalog and advertisement page. (Roy Rogers Museum)

Value $8-12

Schwinn Bicycle advertising booklet. (Roy Rogers Museum)

Value $75-125

Various manufacturing companies advertisements featured in mid-1950s retail merchants full color catalogs. (Roy Rogers Museum)

Value $300-500

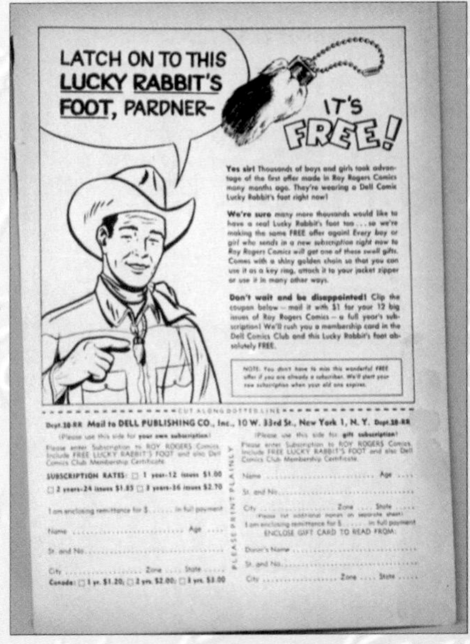

Dell Publishing comic book ad for premium. (R.L.)

Value $3-5

CHAPTER 2
AUTOGRAPHS

Roy Rogers signed his birth name "Len Slye" on some very early and rare photographs before his name changed and he became a movie star. When Republic Pictures changed his name, he signed all his photos and other items with his new name, Roy Rogers, in a very distinct and flowing signature.

Because of his great affection for his horse, Trigger, Roy added Trigger's name to his own when giving autographs. He continued to do so right up until he stopped giving autographs completely in recent years. Dale Evans rarely gave her autograph. You may find a rare photo of both Roy and Dale with both personally signed names. The value of these autographed photos are almost twice as much for a photo with one signature. As far as can be determined, both Roy and Dale signed all of their own autographs without secretarial assistance. Besides photos, prints, and other paper memorabilia, their autographs can also be found on other collectibles, such as Roy Rogers clothing items and vintage toys. Roy's and Dale's autographs have increased in value.

Roy Rogers' and Dale Evans' signatures on a handwritten greeting card or letter may be considered more valuable than on a photograph. The value would depend upon the type of item on which the signature appeared. A personal handwritten letter from Roy Rogers would be valued at $100–250. His signature on a legal document would have a value of $60–125.

The value of Roy Rogers and Dale Evans autographed items varies greatly depending on the item. Also, items that have been personally signed just for you have more value because you now have a dated history of where and when you obtained the autograph. Just how much more they are worth is strictly up to the seller. For personally signed collectibles such as a Roy Rogers hat or cap gun holster, I would add an additional $60–100 to the value of the item. On smaller items such as a trading card, I would add $25–35 to its value. The least valued Roy Rogers or Dale Evans autograph would be a signature on a 3- by 5-inch plain white card with a value of $25–35. These autographs are usually sold matted with an 8- by 10-inch photo of Roy or Dale. Color photographs with one or both of their signatures are worth about 50–100 percent more than autographed black and white photos. Limited edition prints, posters, paintings, and other art objects are some of the most valuable Roy Rogers and Dale Evans autographed items. These items are valued at $200–300 more, depending on the item with their handwritten signature.

Close-up of Roy Rogers and Dale Evans autograph on color photo.

Roy Rogers and Dale Evans autographed 8″ x 10″ black and white photograph.
Value $50-75

George "Gabby" Hayes autograph.
Value $20-30

Autograph on photograph. Note the photograph's original envelope.
Value $45-75

Roy Rogers autographed 8″ x 10″ color photograph.
Value $90-125

Roy Rogers autographed 8″ x 10″ black and white photograph.
Value $50-75

Dale Evans autographed 5″ x 7″ black and white photograph.
Value $25-40

Roy Rogers autographed
8″ x 10″ color photograph.
Value $90-125

Roy Rogers autographed comic.
Value $50-75

Roy Rogers autographed
5″ x 7″ black and white
photograph.
Value $35-50

Roy Rogers autographed
8″ x 10″ color photograph.
Value $90-125

Roy Rogers autographed
8″ x 10″ color photograph.
Value $90-125

Roy Rogers autographed
8″ x 10″ color photograph.
Value $90-125

CHAPTER 3

Badges, Pin-back Buttons and Jewelry

From the late 1940s to the late 1950s, General Foods Company (Quaker Oats) and other companies, including Dell Comics, advertised premium offers on their products. These items consisted of five- and six-point star deputy sheriff badges, a colorful group of pin-back buttons in different sizes, rings, and other items. Post Cereals and Dell Comics ads offered the greatest array of premium products.

Post Cereals Company offered more than 30 different one-piece pin-back buttons. In 1953, General Foods produced Roy Rogers pin-back buttons in association with their sponsorship of Roy's radio and television shows. These buttons depicted the likenesses of Roy Rogers, Dale Evans, Pat Brady, Trigger, Bullet, Nellybelle, and Buttermilk. Other pin-back buttons include illustrations of Roy's guns, boots, and saddle, a yellow Roy Rogers Sheriff's badge, Roy's brand, Dale's brand, and a Roy Rogers Junior Deputy shield badge.

Most pin-back buttons were made in Providence, Rhode Island, and in Long Island City, New York. The buttons ranged in sizes from 3/4 inch diameter to large 1 5/8 inches diameter.

There were some Roy Rogers buttons and Roy Rogers with Trigger buttons manufactured with ribbons and small metal figures attached, such as a steer's head. These buttons with attachments measure 1 3/4 inches long and have a value of $50–65. A jumbo pin-back button put out by Post Grape-Nuts Flakes via "The Roy Rogers Television and Radio Shows" from 1950-1953 depicts a likeness of Roy with lettering "King of the Cowboys" on top. These buttons are valued at $55–75 and are 1 5/8 inches in diameter.

Complete pin-back button sets and jewelry sets have greater value than individual pieces. Republic Pictures issued a variety of pin-back buttons, as did the Roy Rogers and Dale Evans fan clubs. These buttons were produced in white background with black lettering depicting an image of Roy Rogers and produced in the late 1940s. Pin-back buttons manufactured after 1965 and to the present are of a two-piece construction. Most pin-back buttons do not show the manufacturer's name.

Roy Rogers badges
Top: Roy Rogers 2 1/8″ diameter plastic five-point star plastic badge that features a black and white photo of Roy. Very rare. (C.Q.)
Value $85-125

Middle: Roy Rogers Deputy Sheriff six-point badges, gold-plated metal 2 3/4″ diameter. They originally sold for 25¢. (C.Q.)
Value $45-65 each

Bottom: Roy Rogers Deputy base metal 1 3/4″ diameter six-point star badge depicting crossed guns in center in raised relief. (C.Q.)
Value $45-65

Close-up of Roy Rogers Deputy badge showing detail of raised metal design. (C.Q.)

Post Grape-Nuts Flakes ad for Roy Rogers pin-back buttons. Ad found on back of Dell comic book. (R.L.)

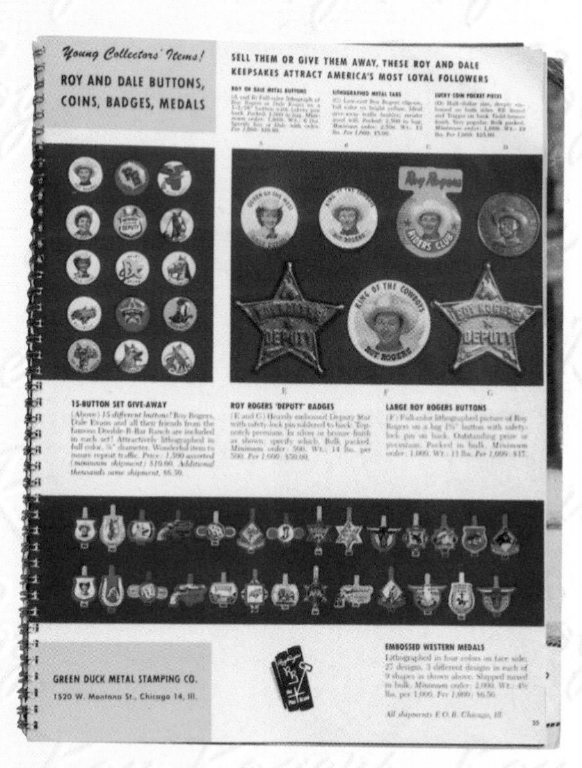

Young Collectors' Items!

ROY AND DALE BUTTONS, COINS, BADGES, MEDALS

SELL THEM OR GIVE THEM AWAY, THESE ROY AND DALE KEEPSAKES ATTRACT AMERICA'S MOST LOYAL FOLLOWERS

15-BUTTON SET GIVE-AWAY

ROY ROGERS 'DEPUTY' BADGES

LARGE ROY ROGERS BUTTONS

EMBOSSED WESTERN MEDALS

GREEN DUCK METAL STAMPING CO.
1520 W. Montana St., Chicago 14, Ill.

Buttons, coins, badges, and medals featured in a one-page retail merchandisers catalog. Items depicted were manufactured by Green Duck Metal Stamping Co., Chicago, IL. (Roy Rogers Museum)

Small Roy Rogers pin-back buttons. (C.Q.)
Value $20-30 each

Roy Rogers large 1 5/8" diameter pin-back button. (C.Q.)
Value $35-60

Roy Rogers pin-back button. (C.Q.)
Value $30-40

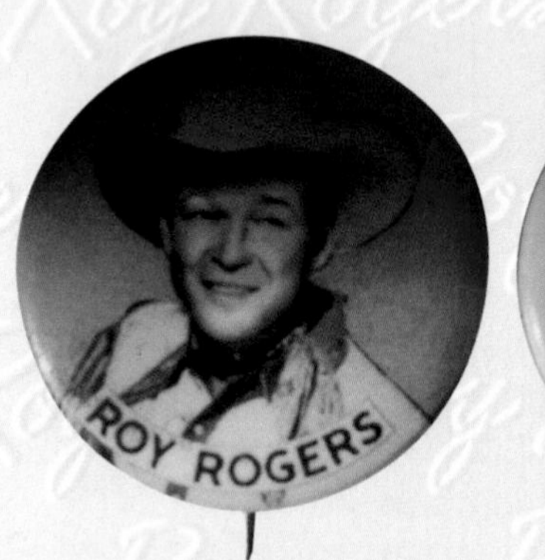

Two Roy Rogers pin-back buttons. (C.Q.)
Value $30-40 each

George "Gabby" Hayes pin-back button. (C.Q.)
Value $25-35

Roy Rogers pin-back button with ribbons and gold-plated boots. (C.Q.)
Value $50-65

Roy Rogers pin-back button with ribbon and pearl-handled jack knife. (C.Q.)
Value $55-70

Roy Rogers pin-back button with ribbon and megaphone. (C.Q.)
Value $55-70

Roy Rogers pin-back button with ribbon and metal hat. (C.Q.)
Value $55-70

Roy Rogers pin-back button with ribbon and moneybag. (C.Q.)
Value $55-70

Roy Rogers sterling silver ring with horseshoe design. (C.Q.)
Value $85-160

Sterling silver ring with raised metal relief depicting Roy Rogers on rearing Trigger sold by Sears. (C.Q.)
Value $150-300

Roy Rogers photo 1950s gold-plated base metal ring. (C.Q.)
Value $75-110

Dale Evans official necklace, white metal. (C.Q.)
Value $40-60

Collection of three Roy Rogers rings. (B.W.)
Value $150-300 each

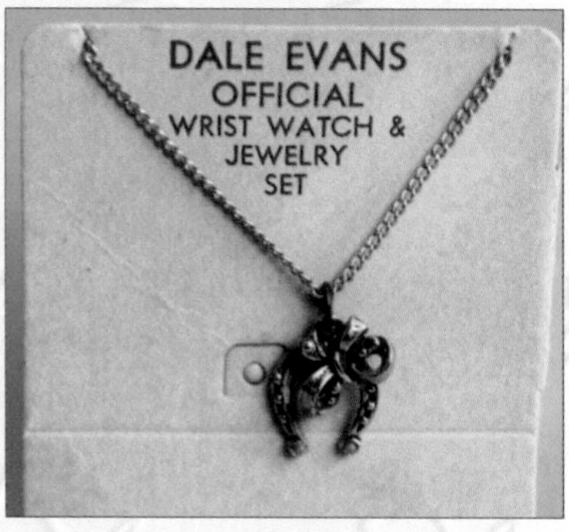

Dale Evans official necklace in horseshoe design with yellow-colored metal. (R.L.)
Value $40-60

Dale Evans bracelet. (M.M.)
Value $60-85

Note: Not shown is a 1950s Roy Rogers charm bracelet, 6" length.
Value $65-90

Roy Rogers gold-plated tie-bar and small belt buckle. (C.Q.)
Value $35-50 each

Roy Rogers and Dale Evans plastic charm from early 1950s. (C.Q.)
Value $30-60

Roy Rogers and Trigger television charms, 1950s. (C.Q.)
Value $35-80 each

Extremely rare Gabby Hayes Quaker Puffed Rice Cereal box with cannon ring ad, premium, and ring. (M.M.)
Value $600-750 complete set as shown
Value $180-250 cannon ring only

Rider's Roy Rogers lucky piece in white metal came in two sizes. (R.L.)
Value $30-65 small; $40-80 large

Collection of Roy Rogers, Gabby Hayes, and Trigger charms, plastic frames, black and white photos, 5/8" x 7/8" 1950s. (R.L.)

Value $30-60 each

Roy Rogers cowboy boot cuff links and saddle tie bar set, gold-washed metal. These sets were distributed at a Roy and Dale Performance in Madison Square Garden in the 1950s. (C.Q.)

Value $40-60

Roy Rogers silver-plated tie bar. (C.Q.)

Value $35-50

Roy Rogers wood ring display stand with sign "These are my favorite rings." Very rare. (B.W.)

Value $125-200

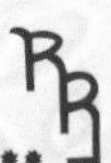

CHAPTER 4

BOOKS

Roy Rogers and Dale Evans biographical and inspirational books provide an intimate look into their family's physical and spiritual lives. These very personal books give great insight to times of "Happy Trails" and times when their lives were extremely difficult to bear. Roy and Dale, both devout Christians, share with us their family life values and their enormous faith in God that helped them through the tragedies of losing three children.

Dale Evans, a very accomplished author, wrote books describing these difficult times and how the love of God, family, and friends helped her and her family. Her books continue to offer great support and help for families that have suffered through similar ordeals and have touched the lives of thousands of people. Through her writings, she and Roy have been true witnesses of their own faith in spreading the word of God and helping others. Some Dale Evans and Roy Rogers books are also available in paperback.

Roy "Dusty" Rogers, Jr., has also authored a wonderful book with Karen Ann Wojahn titled *Growing Up with Roy and Dale*. This popular and humorous book, published by Regal Books, was written in 1973 and reprinted in 1978 and 1984. In this book, Dusty shares his experiences with the reader as the son of "The King of the Cowboys" and "The Queen of the West." I highly recommend it to everyone.

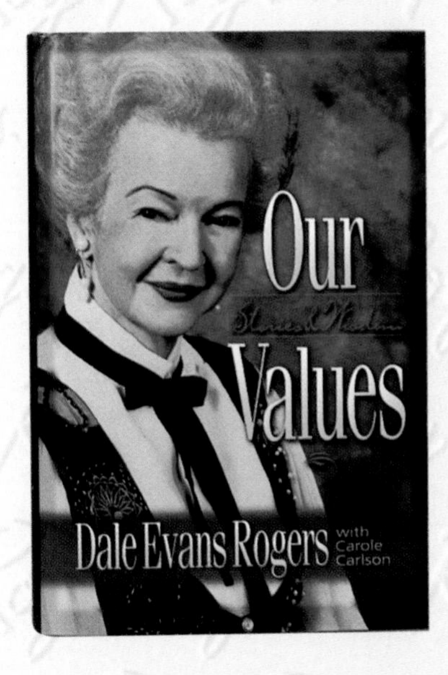

Our Values, by Dale Evans Rogers. Fleming H. Revell Publishers, 1997. (R.L.)
Value $20-30

No Two Ways About It, by Dale Evans Rogers. Fleming H. Revell Publishers, 1963. (R.L.)
Value $20-30

Happy Trails: Our Life Story, by Roy Rogers and Dale Evans with Jane and Michael Stern. Fireside, 1994. (C.Q.)
Value $25-45

The Story of Roy Rogers and Dale Evans, Happy Trails by Roy Rogers and Dale Evans with Carlton Stowers. Guidepost, 1979.(R.L.)
Value $15-20

My Favorite Christmas Story, by Roy Rogers with Frank S. Mead. Revell Publishing, 1960. (R.L.)
Value $20-30

Salute to Sandy, by Dale Evans Rogers. Fleming H. Revell Publishers, 1967. (C.Q.)
Value $20-30

Dale Evans Prayer Book for Children, by Dale Evans. Golden Press, 1956. (R.L.)
Value $20-30

Growing Up with Roy and Dale, by Roy "Dusty" Rogers, Jr. Regal Books, 1973. (C.Q.)
Value $20-30

Other Books Authored by Dale Evans Rogers

Angel Unaware, by Dale Evans Rogers. Fleming H. Revell Publishers, 1958.
Value $15–25

Dearest Debbie, by Dale Evans Rogers. Fleming H. Revell Publishers, 1958.
Value $15–25

Christmas Always, by Dale Evans Rogers. Fleming H. Revell Publishers, 1958.
Value $15–20

Dale: My Personal Picture Album, by Dale Evans Rogers. Fleming H. Revell Publishers, 1971.
Value $20–25

In the Hands of the Potter, by Dale Evans Rogers. Thomas Nelson Publishers, 1994.
Value $25–35

Let Us Love, by Dale Evans Rogers. Words Books Publishers, 1982.
Value $20–25

My Spiritual Diary, by Dale Evans Rogers. Fleming H. Revell Publishers, 1955.
Value $20–25

Only One Star, by Dale Evans Rogers. Words Books Publishers, 1988.
Value $25-30

Time Out Ladies! by Dale Evans Rogers. Fleming H. Revell Publishers, 1966.
Value $15–20

To My Son, by Dale Evans Rogers. Fleming H. Revell Publishers, 1957.
Value $20–25

Trials, Tears and Triumph, by Dale Evans Rogers. Fleming H. Revell Publishers, 1977.
Value $20–25

The Women at the Well, by Dale Evans Rogers. Fleming H. Revell Publishers, 1970.
Value $20–25

Where He Leads, by Dale Evans Rogers. Pillar Books Publishers, 1971.
Value $7–10

The Answer is God,
by Elise Millar Davis.
McGraw-Hill Publishers,
1955. (R.L.)
Value $25-35

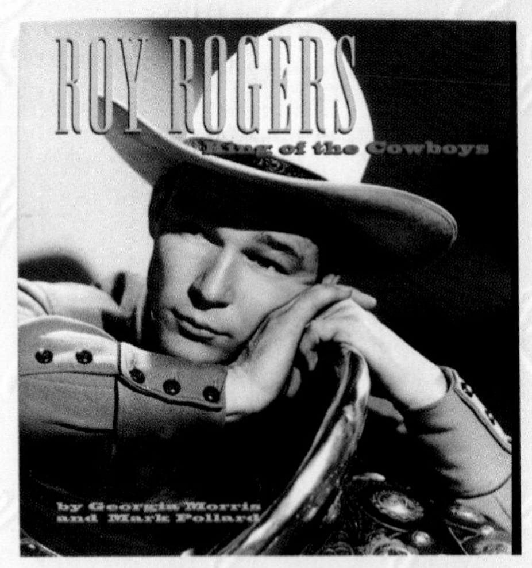

Roy Rogers: King of the Cowboys,
by Georgia Morris and Mark Pollard.
Collins Publishers, 1994. (R.L.)
Value $30-40

Trigger Remembered,
by William Witney.
Earl Blair Enterprises,
1989. (C.Q.)
Value $25-35

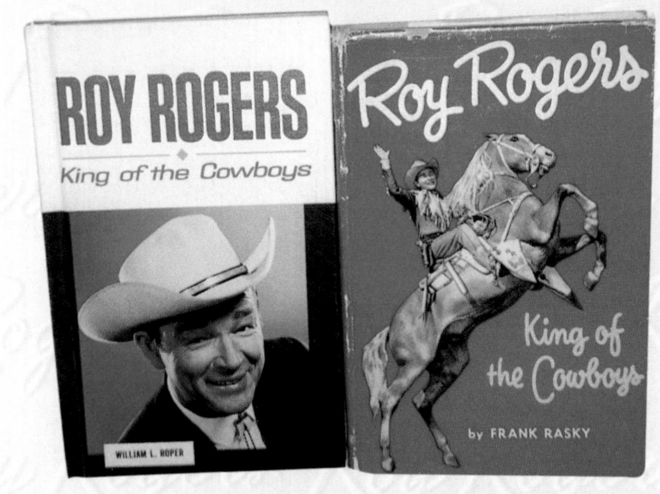

Roy Rogers: King of the Cowboys, by William L. Roper. T.S. Denison & Co., 1971. Very rare. (R.L.)
Value $45-75

Roy Rogers: King of the Cowboys, by Frank Rasky. Julian Messner Company Publishers, 1958. (R.L.)
Value $45-75

Other Books About Roy Rogers and Dale Evans

The Angel Spreads Her Wings, by Maxine Garrison. McFarland and Company, Inc. Publishers, 1995.
Value $15–20

The Roy Rogers Book, by David Rothel. Empire Publishing Co., 1987.
Value $25–35

Roy Rogers, by Robert W. Phillips. McFarland & Co., Inc. Publishers, 1955.
Value $45–60

CHAPTER 5

Cap Guns-Holsters-Knives

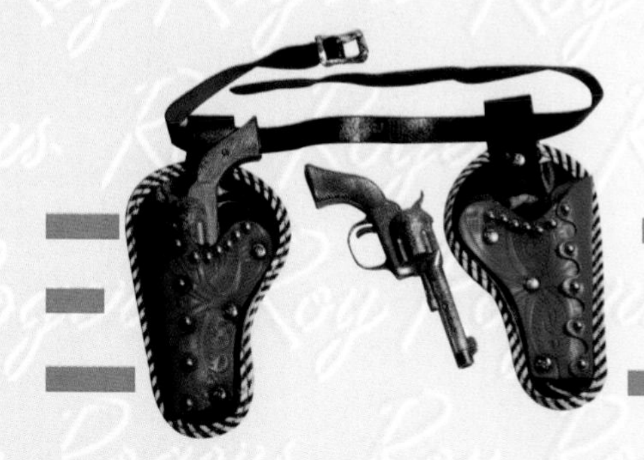

Cap guns and holsters are two of the most popular and sought after western hero collectibles. In today's market, they are valued at higher prices than real handguns and holsters currently being manufactured. A cap gun with a western hero's name on it is valued at two to three times as much as the same cap gun without a name. One of the rarest and most valued cap guns is the cast iron Roy Rogers Long Tom made by Kilgore in the late 1930s and early 1940s. The price for one of these well-made guns in excellent condition is about $1,200.

With the enormous popularity of western movies and television shows from the 1940s through the 1960s, toy companies such as Hubley flooded the market with every conceivable type of cap gun and holster. Guns ranged in sizes from 2 1/2" to 13" long. Most of the cap guns manufactured were extremely realistic and looked very much like the real guns. Cap guns like the Nichols Stallion 45 have about the same weight and feel of a real Colt .45. This twelve-inch long cap gun was a real handful for a young buckaroo to handle. It would seem that many of these larger cap guns were made more for adults than for children because of their highly detailed engraving and fancy raised relief metal and plastic grips. Some have revolving cylinders that took one and two-piece metal toy bullets.

Roy Rogers and Dale Evans cap guns were manufactured by George Schmidt, Classy Products, Leslie Henry, Lone Star, Buzz Henry, Stevens, Hubley, Kilgore, and Balantyne. Many of these cap guns were assembled with nuts, bolts, and screws, which makes it less difficult to restore than the ones put together with rivets. These guns also contained hammer springs, trigger springs, and barrel release springs, almost like a real handgun. Parts for restoring cap guns are still being made by some very talented and devoted individuals like Frank McBath and Herb Taylor. Gold-plated cap guns have a higher value than the same gun with a nickel finish. Also, cap guns with a dummy hammer are more valuable than guns with a ridged hammer. Dummy hammers are smooth fat metal hammers that will not fire a cap.

Cap guns with dummy hammers were made to sell in certain states and cities that prohibited the use of fireworks and caps. These cap guns are rarer than the cap-firing guns and can generally be found in nicer condition, as they will not have cap corrosion in the hammer area.

Original boxes for cap guns, holsters, and other collectibles greatly add to the item's value. A very rare box can be worth as much as the item. Be sure to examine the box carefully before purchasing, as reproduction boxes are currently being produced on laser copying equipment.

Roy Rogers and Dale Evans holsters were some of the fanciest holsters produced. They were made of leather and elaborately decorated with metal stamped Roy Rogers studs, different colored plastic jewelry, raised relief metal conches, plastic and wood bullets, cut-outs, and fringe. Generally, the fancier and larger double holsters command higher prices than smaller and less fancy ones. Some very elaborate Roy Rogers holsters have gold or silver-colored foil embossed over the leather. Holsters of this type with all the metal foil still intact are very rare and highly valued because they were made in limited production numbers. Dale Evans cap guns are rarer and more valuable than Roy Rogers rigs because fewer were made and they are in higher demand.

Classy, Keyston Brothers, Halco, Leslie Henry and other little known manufacturers made Roy and Dale leather holsters. Like cap guns, expert leather craftsmen can restore leather holsters to look new. Two of the finest leather restoration experts that I highly recommend are John Coatney and Joe Wishart. Restored Roy Rogers and Dale Evans cap guns can look better than new and have almost as much value as unrestored mint condition pieces.

CAP GUN HOLSTER ADS

Roy Rogers 21–piece western outfit ad for retail stores.
(Roy Rogers Museum)

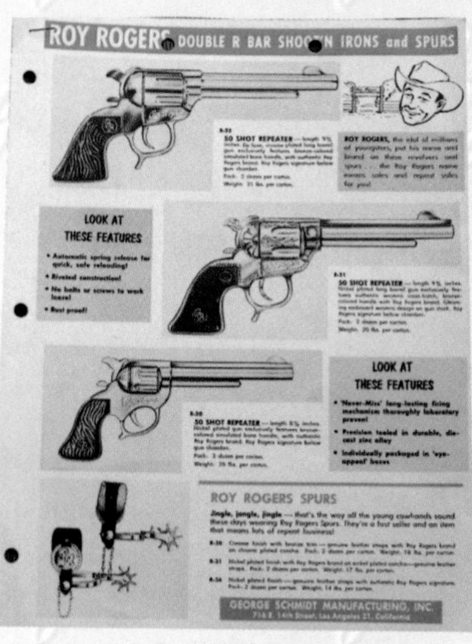

George Schmidt Manufacturing Company catalog ad for Roy Rogers cap guns.
(Roy Rogers Museum)

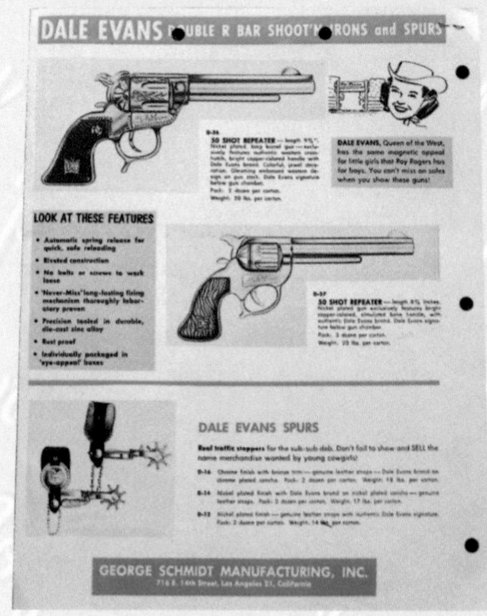

George Schmidt Manufacturing Company retail merchants catalog ad for Dale Evans cap guns.
(Roy Rogers Museum)

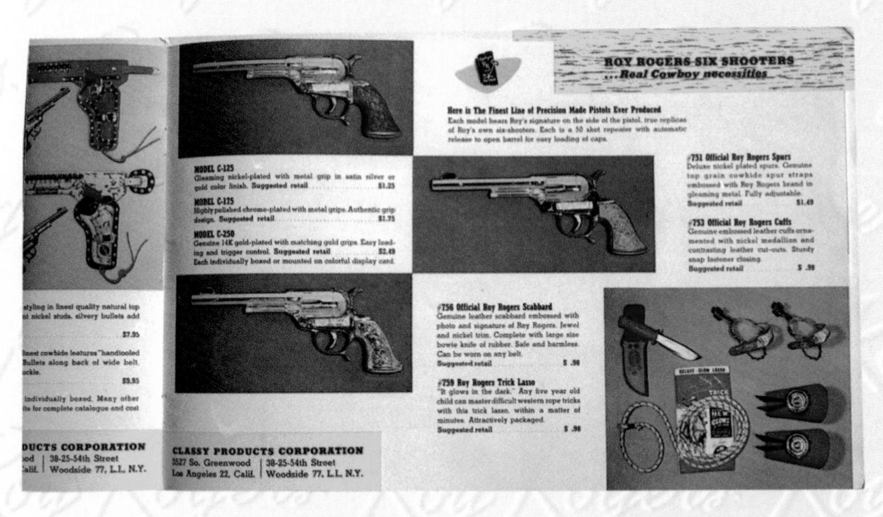

Classy Products Corporation retail merchants catalog advertisements for Roy Rogers cap guns and holsters.
(Roy Rogers Museum)

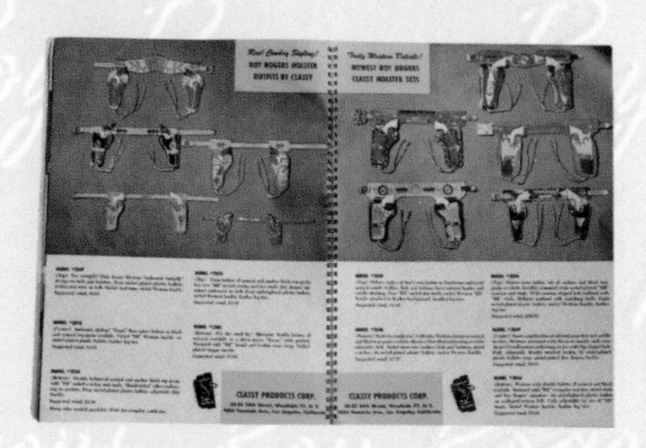

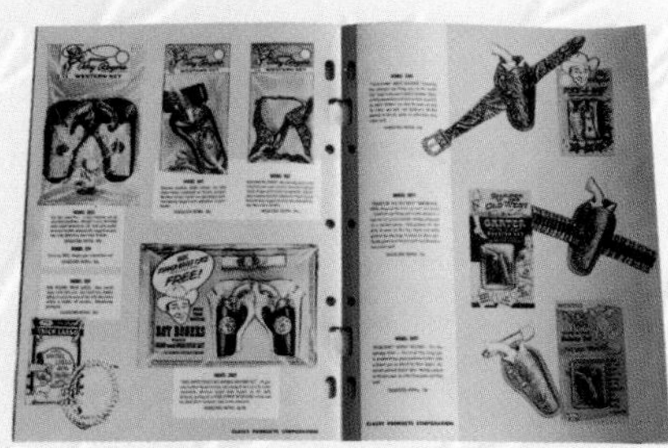

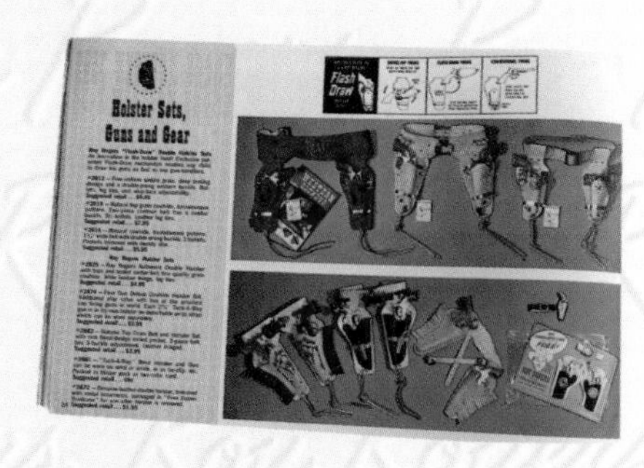

Classy Products Corporation retail merchants catalog advertisements for Roy Rogers cap guns and holsters. (Roy Rogers Museum)

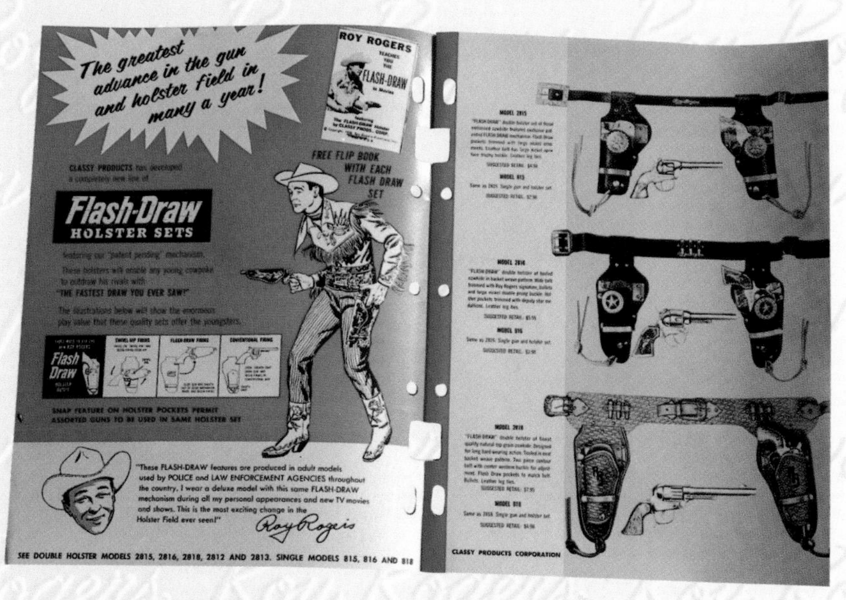

Classy Products Corporation retail merchants catalog advertisements for Roy Rogers cap guns and holsters. (Roy Rogers Museum)

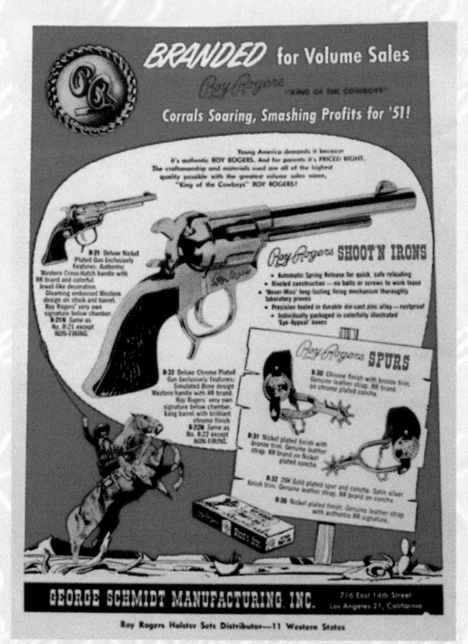

George Schmidt Manufacturing Company magazine ad for cap guns and spurs. (Roy Rogers Museum)

Value $20-30

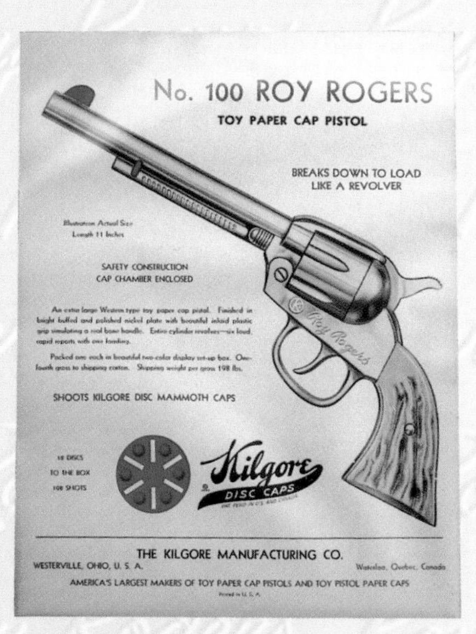

Kilgore Manufacturing Company display card for Roy Rogers cap guns. (C.Q.)

Value $25-40

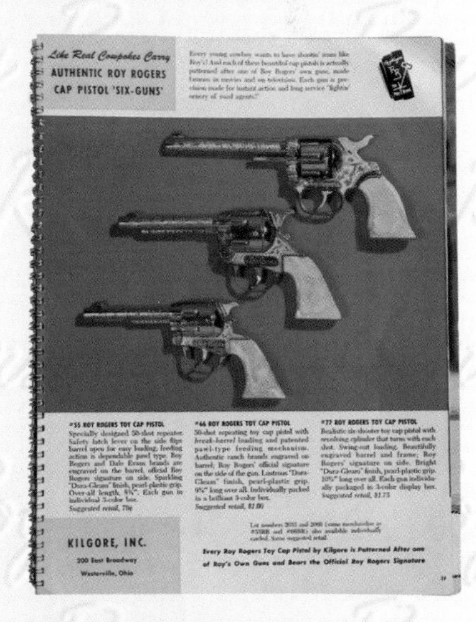

Kilgore Manufacturing Company retail merchants catalog ad for three different Roy Rogers cap guns. (Roy Rogers Museum)

BUZZ HENRY CAP GUNS

Roy Rogers Buzz Henry rare gold-plated cap gun with white insert plastic grips, 7 1/2" long, 1950-1960. (B.W.)

Value $125-200; Box only value $75-125

Dale Evans nickel-plated cap gun with red plastic insert grips, very rare, 1950-1955. (G.S.)

Value $225-350; Box only value $200-350

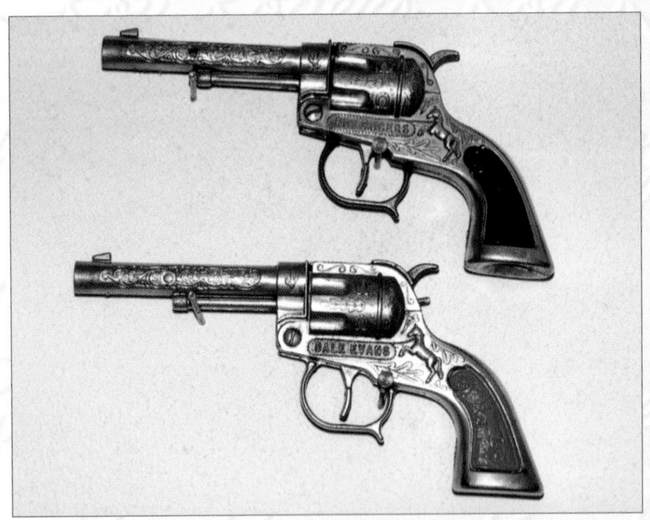

Dale Evans Buzz Henry nickel-plated cap guns with red and black plastic insert grips. Rare cowgirl guns. (B.W.)

Value $200-350 each

Dale Evans Buzz Henry very rare gold-plated cap guns with black plastic insert grips. (C.Q.)

Value $300-450 each

Roy Rogers Buzz Henry 7 1/2" long cap gun with dull finish and black insert grips, 1950-1960. (C.Q.)

Value $100-165

Roy Rogers Buzz Henry cap gun in bright nickel finish and black insert grips, 1950-1960. (C.Q.)

Value $115-175

Roy Rogers Buzz Henry cap gun with nickel finish and white grips, 1950-1960. (C.Q.)

Value $115-165

CLASSY CAP GUNS

Between 1950 and 1960, Classy Products Company distributed and sold several different sized cap guns. Lonestar and B.C.M manufactured many of these guns in England. The smallest Classy cap guns are models R-20 and C-31. Both of these single-shot guns are 2 3/4" long and labeled "Tuck-A-Way Gun." The single-shot "Pee Wee" Model R-30 is 5 1/2" long. The Model R-50 is 8" long and the Model R-90 is a 10" repeater that fires a 50-shot roll of caps.

Model R-100 is one of the largest Classy cap guns measuring 11" and made with a variety of metal scroll, floral, and stag raised relief grips. It was also manufactured with a chrome-plated finish and a rare 14kt gold-plated finish. The rarest of all Classy cap guns if the Model R-202 in a gold-plated finish. The gun was also available in a chrome-plated finish. The R-202 has a revolving cylinder, used small 2-piece bullets, and has impressive white plastic grips with a raised relief image of an Indian Chief.

Very rare pair of Model R-202 gold-plated 10 1/2" engraved frame cap guns. (C.Q.)

Value $400-800 each

Model R-100 rare pair of nickel-plated frame guns with gold-plated barrel, cylinder, and scroll grips. (C.Q.)

Value $280-450 each

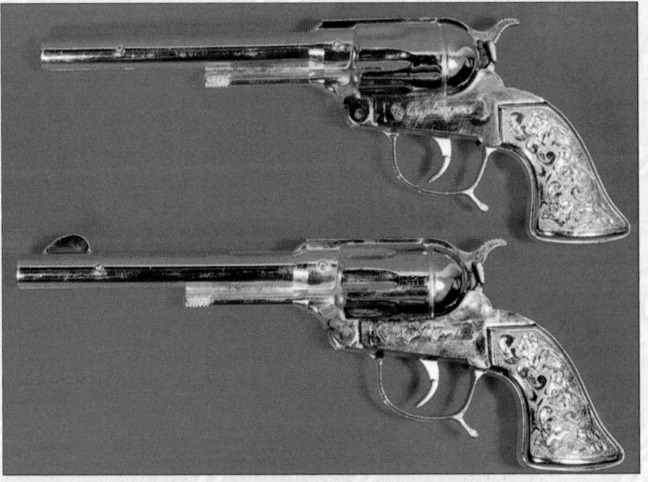

Very rare matched pair of Model R-100 gold-plated guns with metal grips. (C.Q.)

Value $400-800 each

Close-up showing ornate detail of floral scroll grips. (C.Q.)

Model R-100 with nickel finish and stag grips. (C.Q.)

Value $200-350

Polished nickel cap gun with pewter gray grips. (C.Q.)

Value $175-350

Model R-60 matched pair with rare gold-plated finish and rearing horse grips. (C.Q.)

Value $200-300 each

Model R-60 cap guns with engraved nickel-plated finish. (C.Q.)

Value $150-250 each

Model R-30 Pee Wee 5 1/2″ pair of all metal single shot cap guns in dull finish. (C.Q.)

Value $45-100 each

Model R-90 cap gun in polished gold finish with copper-plated rearing horse grips. (C.Q.)

Value $175-325

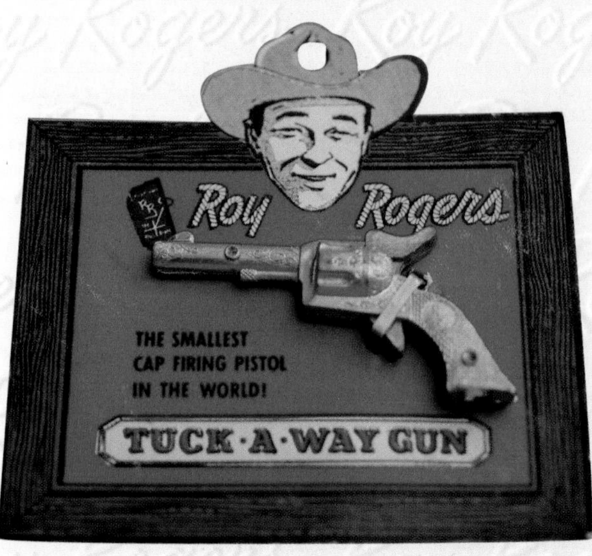

Model R-20 single shot 2 3/4″ all metal "Tuck-A-Way Gun" on card. (C.Q.)

Value $45-110

Model R-20 all metal gold-finished single shot cap gun with Holster. (C.Q.)

Value $60-130

Combination western outfit set with Model R-60 cap guns and box. (R.L.)

Value $450-750

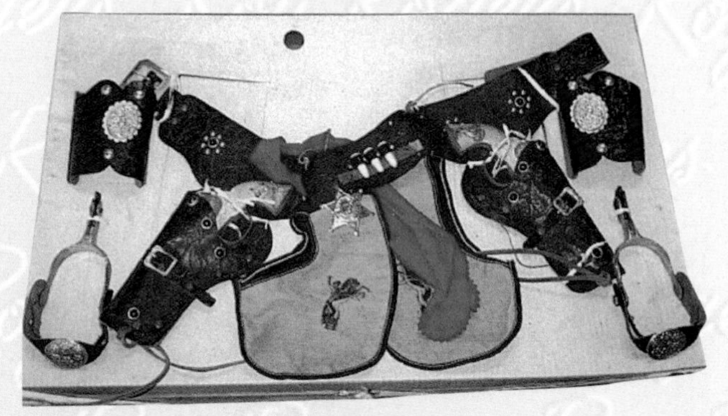

Combination western outfit set with Model R-60 cap guns and box. (R.L.)

Value $450-750

Model R-30 single shot cap gun with holster in original display bag. (C.Q.)

Value $75-125 as shown

KILGORE CAP GUNS

Very rare Roy Rogers Kilgore cast iron cap gun with polished finish, long top strap and riveted cream-colored stag grips, 10 1/2", red plastic front sight and ejection lever, made in late 1930s. Long Tom or Roy Rogers name imprinted in red letters on grips. (C.Q.)

Value $900-1,500 gun only

Value $300-500 Kilgore pistol box

Value $1,000-1,600 "Long Tom" Model with etched side plate

Pair of very rare cast iron "Big Horn" cap guns with two-piece cylinders and Roy Rogers name lightly etched on side of frame, very natural-looking cream and brown-colored grips. (C.Q.)

Value $850-1,350 each

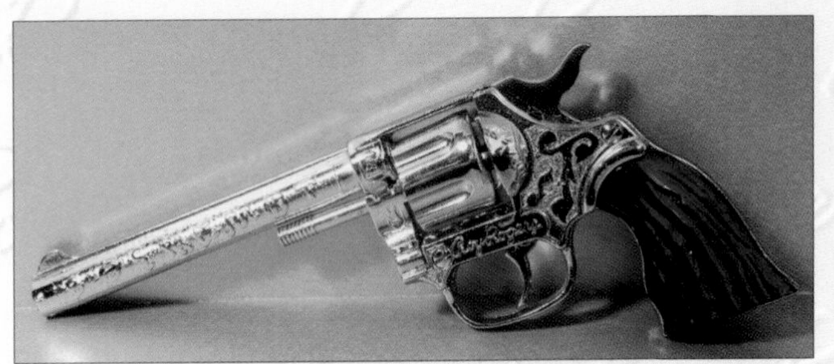

Very rare large 10 1/4" gold-plated fancy engraved cap gun with revolving cylinder and black horse head plastic grips. (R.L.)

Value $300-400

Rare 1955–1960 large six-shot revolving cylinder 10 1/4" cap gun with high polished finish and white horse head grips. (C.Q.)

Value $200-350

1938-40 very rare 9" cast iron "Big Horn" cap gun with reddish brown stag grips. (C.Q.)

Value $900-1,400

Die cast gold-plated rare 9" cap gun with dark brown swirl horse head grips. (C.Q.)

Value $175-275

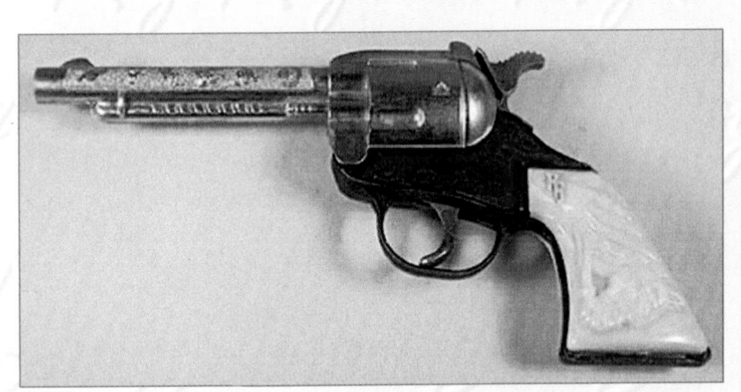

Die cast 9" very rare bronze frame and nickel finished barrel and cylinder cap gun with white raised relief horse head plastic grips, 1955. (B.W.)

Value $150-250

Rare 1955 nickel die cast 9" cap gun with chocolate brown swirl horse head grips. (R.L.)

Value $125-200

1955 dull finish 9" cap gun with cream-colored horse head plastic grips. (C.Q.)

Value $100-175

First model in polished nickel finish, very simple scroll engraving and heavy frame. (C.Q.)

Value $175-275

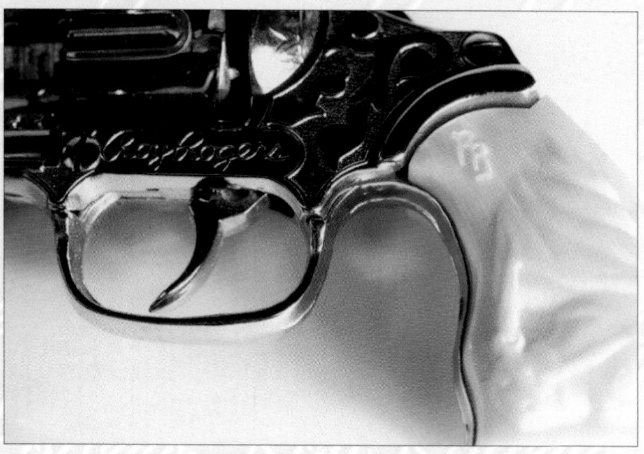

Close-up view of Kilgore's revolving cylinder cap gun, a very beautiful gun with fine detail.

1955 dull finish 9" cap gun on original display card. (C.Q.)
Value $125-200 gun and card

LESLIE HENRY CAP GUNS

First model heavy cap gun in high polished finish and red plastic horse head grips. (C.Q.)
Value $150-275

Second model 1950-1960 cap gun with butterscotch-colored plastic grips. This is the rare "dummy hammer" model. (C.Q.)
Value $200-350

Pair of second model cap guns with gold-plated finish and black grips. These are the smoker models that have little holes in the cap anvil. (R.L.)
Value $225-375 each

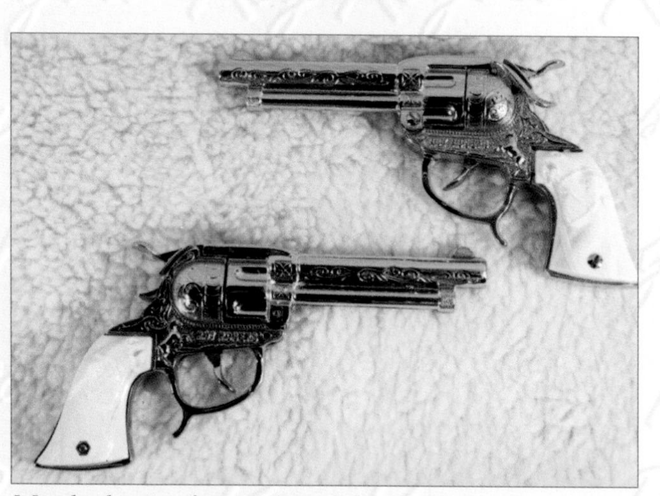

Matched pair of second model cap guns with polished nickel finish and white grips. (C.Q.)
Value $150-275 each

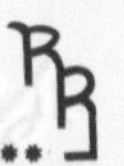

First model 9″ cap gun with high hammer and white horse head plastic grips. (C.Q.)

Value $175-275; Box only value $125-200

Second model 9″ cap gun with dull finish and black plastic grips. (C.Q.)

Value $150-250

Close-up view of Leslie Henry second model 9″ cap gun. (C.Q.)

GEORGE SCHMIDT CAP GUNS

In the 1950s George Schmidt Manufacturing, Inc., made three basic models of cap guns with variations on each model. The three models are Model R-20, Model R-21 and Model R-22. All of these Roy Rogers fifty-shot repeaters were called "shoot 'n irons." Models R-20 and R-21 were nickel-plated with bronze-colored grips. Model R-22, a 9 3/4″ long cap gun was chrome-plated with copper-colored stag grips. The Model R-20 at 8 3/4″ is the smallest of the three. This cap gun has a smooth nickel-plated finish. Models R-21 and R-22 were made with white-painted or copper stag and checkered grips. The Roy Rogers brand symbol in raised relief is found at the top of the grips. Some grip variations were made with red, blue, green, or yellow jewels inserted in this area of the grip. The jeweled grip guns are very scarce and valued higher than non-jeweled grip cap guns. Many of the cap guns manufactured had bronze-colored hammers or triggers. The heavily engraved guns have some engraving on the hammers as well. Engraved George Schmidt cap guns are considered more valuable than the non-engraved models.

Nickel-plated, non-engraved cap gun with red jewel insert white grips, rare. (C.Q.)

Value $250-450

1950-1960 small frame 9″ cap guns with copper-plated hammers, release levers, and stag metal grips. (C.Q.)

Value $200-300 each

Chrome-plated smooth finish cap gun with copper stag grips showing raised relief circular Roy Rogers brand, and name on both sides of this 9 3/4″ gun. (C.Q.)

Value $175-325

Close-up view of George Schmidt cap gun showing Trigger symbol with horseshoe near top of grip.

8 1/2″ cap gun with dull black finish and white grips. (C.Q.)

Value $75-115

Nickel-plated engraved 9 3/4″ long cap guns with copper-plated hammers, triggers, release levers, and checkered grips. This pair has non-engraved ribbed barrels. (C.Q.)

Value $175-325 each

Roy Rogers George Schmidt 9 3/4″ cap gun with fully engraved barrel and frame, nickel-plated release lever and engraved hammer. Copper-plated checkered grips contain blue jewel. (C.Q.)

Value $250-450

Extremely rare pair of Dale Evans nickel-plated guns with ribbed barrels, engraved frame, and red jewel checkered grips shown with their original boxes. (B.W.)

Value $300-500 each for cap guns

Value $250-450 for box only

Two variations of George Schmidt cap guns and the box that one came in. (G.S.)

Value $200-350 red-jeweled engraved gun

Value $150-300 non-engraved gun

Value $100-175 Shoot'n Iron box

LONE STAR CAP GUNS

Lonestar and a few other English manufacturing companies including B.C.M. made cap guns that were sold in England under their own names as well as cap guns that were sold by Classy Products in the United States. Lonestar manufactured several types of large revolving cylinder cap guns that used small 2-piece brass and lead toy bullets. The Roy Rogers Classy Model R-202 is one of Lonestar's rarer models.

England, 8 1/2" cap gun with silver painted finish, lanyard ring, revolving cylinder, and red plastic star in black steer head grips. Another extremely rare cap gun. This one was purchased from an Australian collector. (C.Q.)

Value $300-500

England, 1950s nickel die-cast 9" cap gun with revolving cylinder, engraved frame, lanyard ring, and white steer head grips. Roy Rogers metal medallion insert in grips. This is an extremely rare cap gun. (B.W.)

Value $300-500

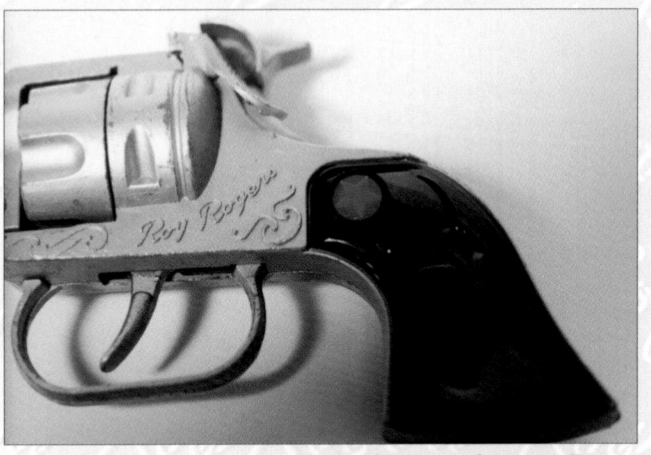

Roy Rogers mystery cap gun close-up showing Roy Rogers in raised relief on side of frame. (C.Q.)

OTHER ROY ROGERS GUNS

Ideal Toy Corporation's "Quick Shooter" Roy Rogers cowboy hat with original box and instructions, 1961. Hat contains a small black double-barreled derringer that pops out when release button on the side of the hat is pushed. (C.Q.)

Value $150-225 hat only

Value $150-225 box only

Roy Rogers flashlight/whistle gun by Charmore Company. Red 3" plastic (also came in black plastic) gun with Roy Rogers name on barrel. Shown on original card. (C.Q.)

Value $35-50 gun only

Value $50-90 with card

Roy Rogers Six-Shooter made by Benlyn Gamecraft, Ltd. This 7" x 10" wood rubber band gun holds up to 32 rubber bands to shoot. Shown in original box, rare. (C.Q.)

Value $75-150

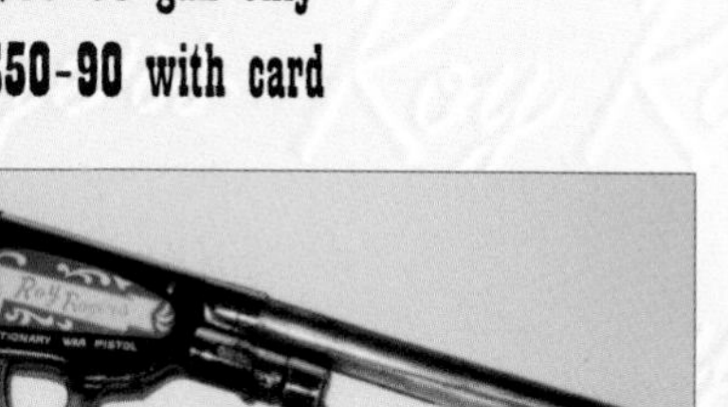

Roy Rogers Japanese gun, very rare, 1950-1960s cork gun. Gun is marked "Roy Rogers Revolutionary War Pistol." (C.Q.)

Value $90-150

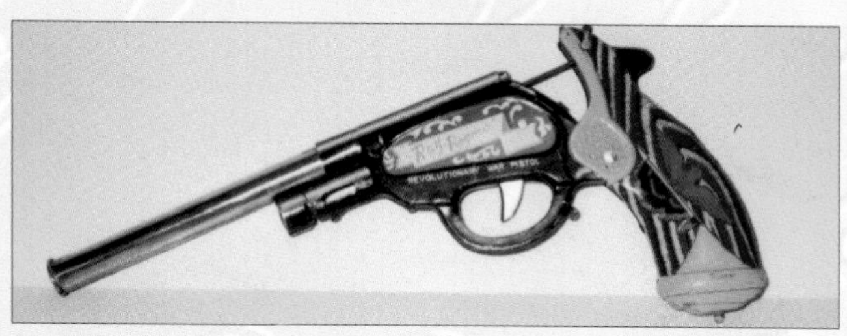

Roy Rogers Japanese cork gun in open position

Roy Rogers Crackin' Good Gun, premium for Post Cereals, 8" long cardboard. (C.Q.)

Value $45-75

Roy Rogers Signal Gun, complete set. (M.M.)

Value $300

Roy Rogers Riders Signal Gun made by Langson Manufacturing Company with original box 5 1/2" x 4". Metal gun is battery-operated. (C.Q.)

Value $100-150 gun only

Value $100-150 box only

CAP GUN BOXES

Roy Rogers Kilgore "Shootin' Iron" box for revolving cylinder cap gun. (C.Q.)

Value $135-200

Roy Rogers Leslie Henry blue and yellow early 1950s box. (C.Q.)

Value $115-150

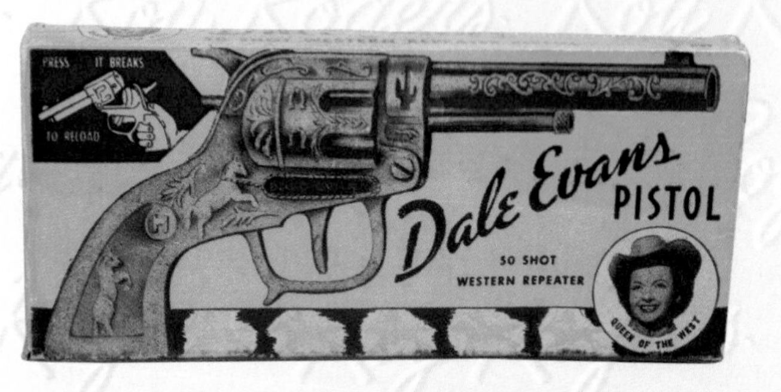

Dale Evans Buzz Henry cap gun box, very rare. (C.Q.)

Value $250-450

Roy Rogers Kilgore "Shootin' Iron" cap gun box. Gun originally sold for 89¢. (C.Q.)

Value $110-180

Roy Rogers Leslie Henry box with pistol that came in it. (R.L)

Value $285-485

Roy Rogers Kilgore two-piece box for the 1938-1940 cast iron cap gun. Most desired and valuable pistol box. (B.W.)

Value $350-550

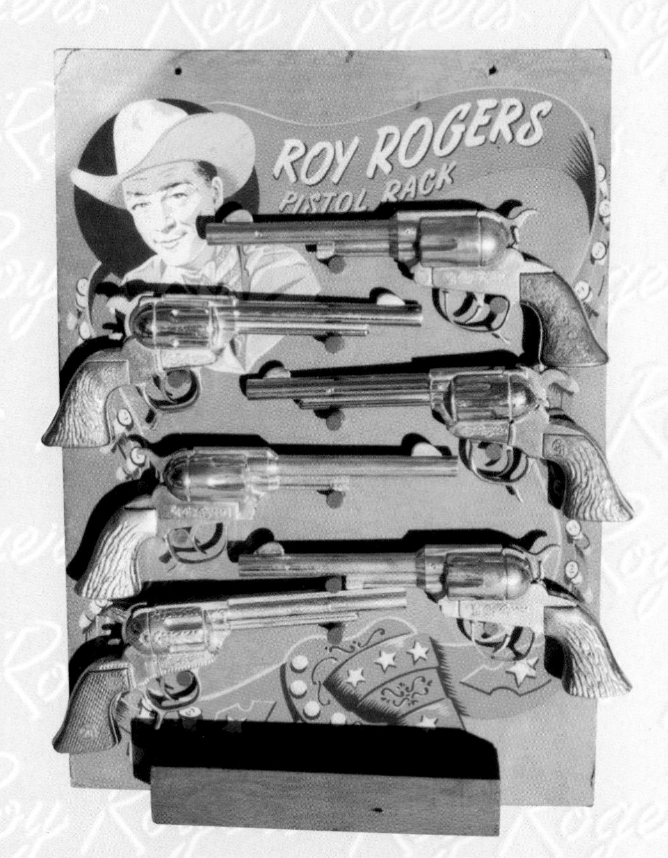

Roy Rogers pistol rack made to hang on wall, very rare. (B.W.)

Value $120-220 rack only

Roy Rogers Kilgore two-piece extremely rare box with blue insert. The mint condition cast iron "Long Tom" pistol is shown in box. This is a beauty! (C.Q.)

Value $350-550 box only

ROY ROGERS HOLSTERS

Roy Rogers putting double set on his son Roy "Dusty" Rogers, Jr. original photo. (Roy Rogers Museum)

Dusty and his sister wearing their Roy Rogers double holster and gun rigs. (Roy Rogers Museum)

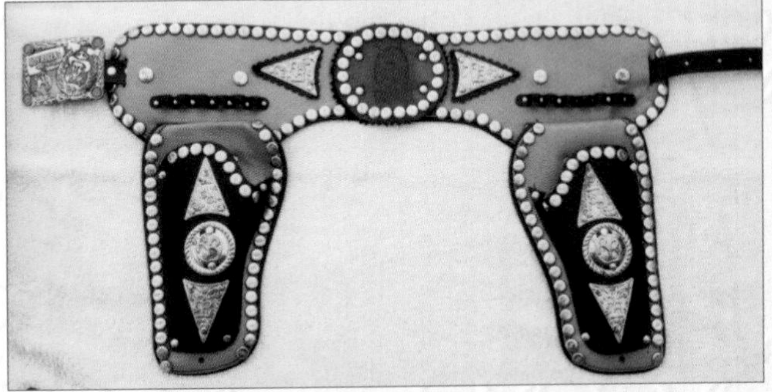

Classy Tan and black holster with many studs and large rodeo buckle, large extra fancy holster has six triangle medallions and Roy Rogers conches on holster pockets. (R.L.)

Value $250-450

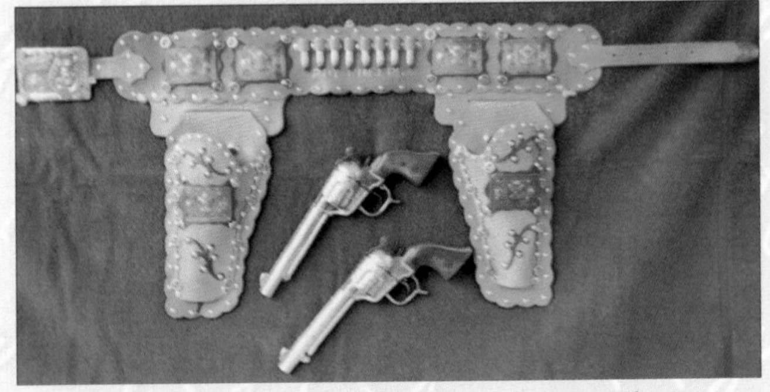

Large double holster, extremely fancy with six gold-plated medallions, large rodeo buckle, many small studs, and Roy's name in metal letters in the middle of the belt. The three-inch wide belt holds eight bullets. Very rare holster with 40 yellow jewels. (R.L.)

Value $450-900 holster without guns

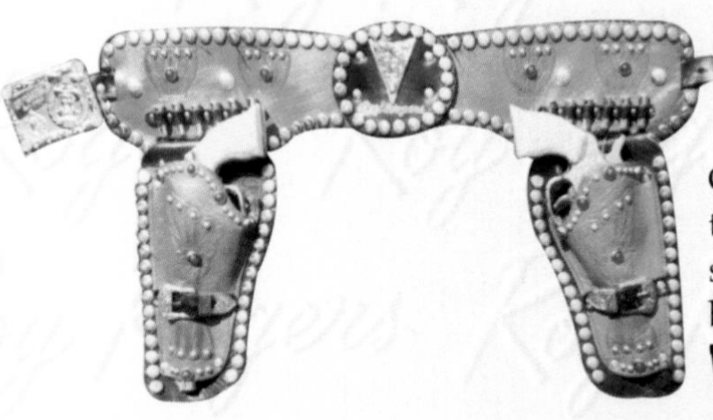

Classy large brown holster with black trim, rodeo buckle, many Roy Rogers studs, buckles on holster pockets, bullet loops, and red jewels. (B.W.)

Value $250-450 holster only

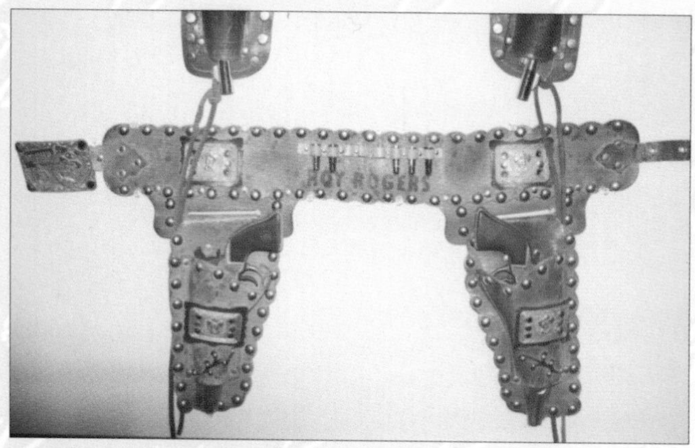

Double tan holster with jeweled rodeo buckle, jeweled medallions and 1/2" studs. (B.W.)

Value $350-600 holster only

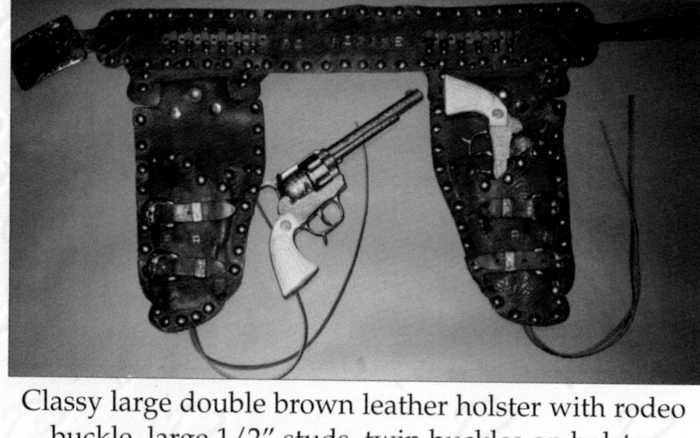

Classy large double brown leather holster with rodeo buckle, large 1/2" studs, twin buckles on holster pockets, and Roy's name in metal letters on middle of belt. (G.S.)

Value $250-400 holster only

Classy large brown and black tooled holster with Roy Rogers rodeo buckle, holster pocket buckles and 16 bullet loops. (C.Q.)

Value $225-375

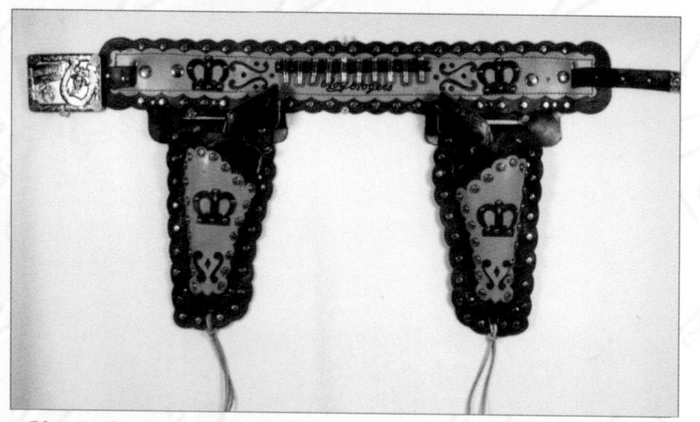

Classy large brown and tan holster, very fancy crown designs and rodeo buckle. (C.Q.)

Value $250-425 holster only

FANCY FOIL OVERLAY HOLSTERS

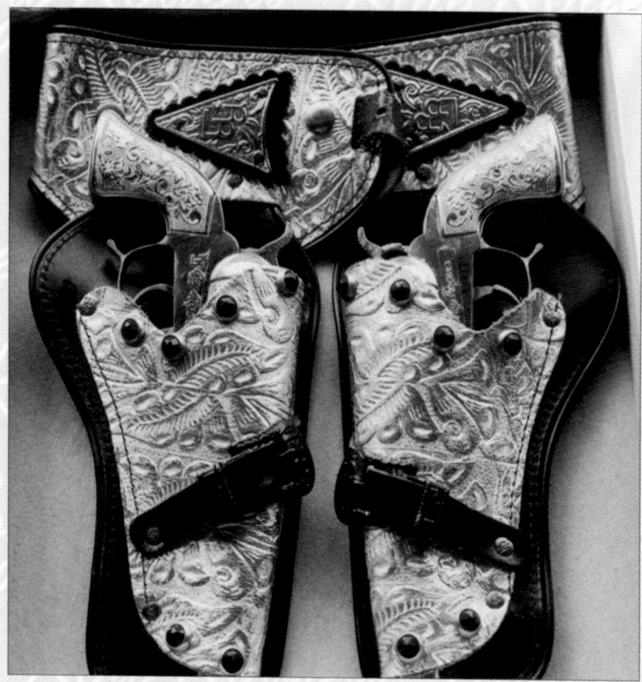

Classy very rare double holster fun set in original mint condition with two-piece box. This is Dusty's personal set from his collection.
(Roy Rogers Museum)

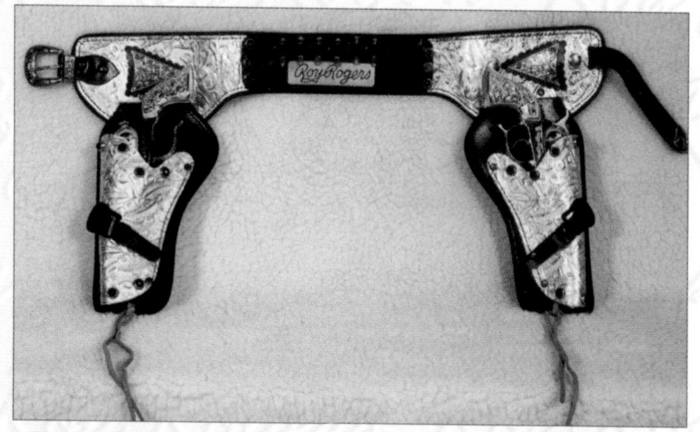

Classy fancy and very rare holster with gold-colored foil overlay on embossed black leather. This black jeweled and gold-plated medallions and buckles holster had very limited production and was given as a prize in movie theater contests. (C.Q.)

Value $350-700 holster only

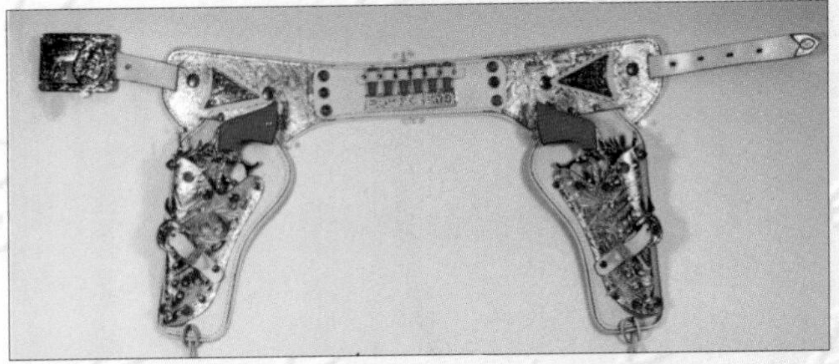

Classy very fancy and rare girls double holster with gold-plated studs, medallions, and buckles. Holster has silver-foil overlay on embossed leather, buckles on holster pockets and red jewels. Roy's name on white belt beneath bullet loops. (C.Q.)

Value $350-650 holster only

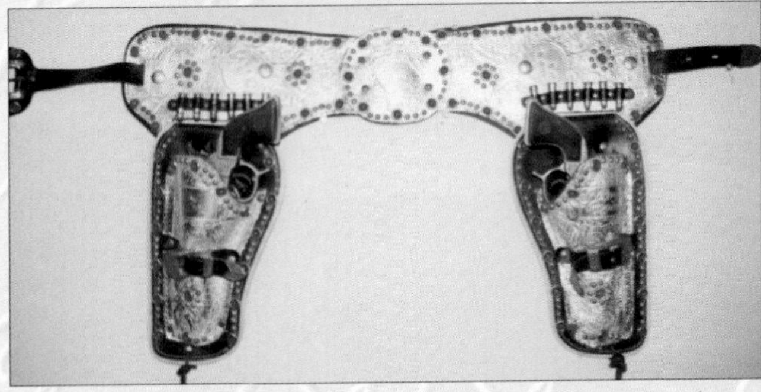

Classy fancy rare holster with silver-foil overlay, lots of red jewels, gold-colored studs and buckles, and black leather trim. (B.W.)

Value $350-650 holster only

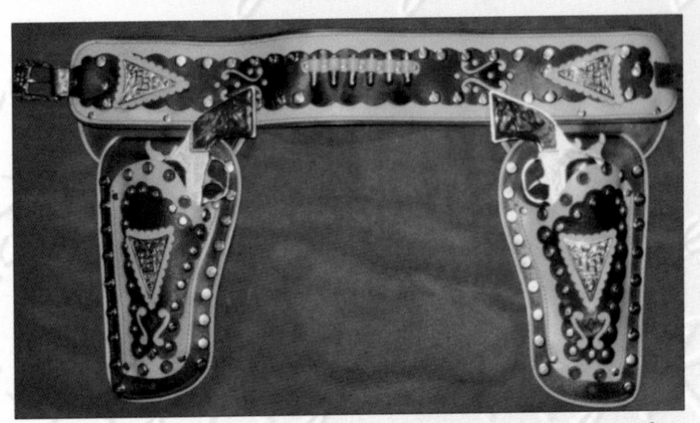

Very fancy large tan and brown holster with triangular Roy Rogers medallions, red jewels, and Roy Rogers studs. (B.W.)

Value $200-350 holster only

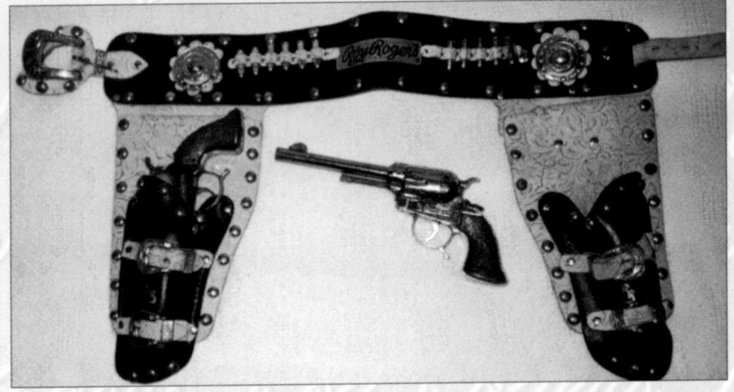

Classy large and fancy black and white tooled leather holster with gold-colored studs, buckle, and holster pocket buckles. Roy's name in the center of the belt, twelve bullet loops with two Roy Rogers medallions. (G.S.)

Value $185-300 holster only

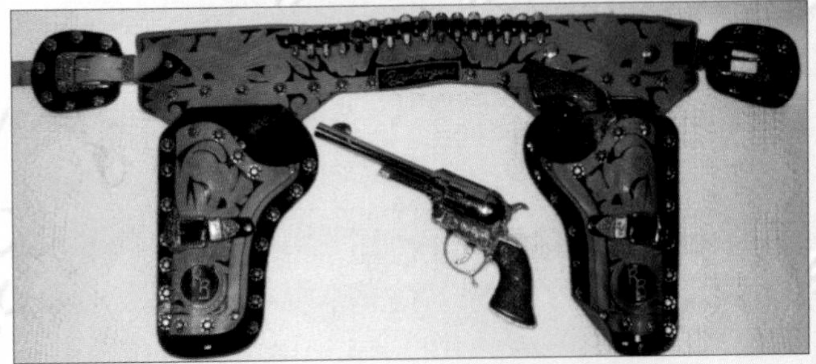

Holster, tooled black and tan leather. Roy Rogers holster pocket buckles, floral-type studs, and double metal leather combination buckles, 1955. (G.S.)

Value $200-300 holster only

Holster, embossed medallion and dark brown leather, very small studs, and floral-type studs, 1950s. (G.S.)

Value $185-300 holster only

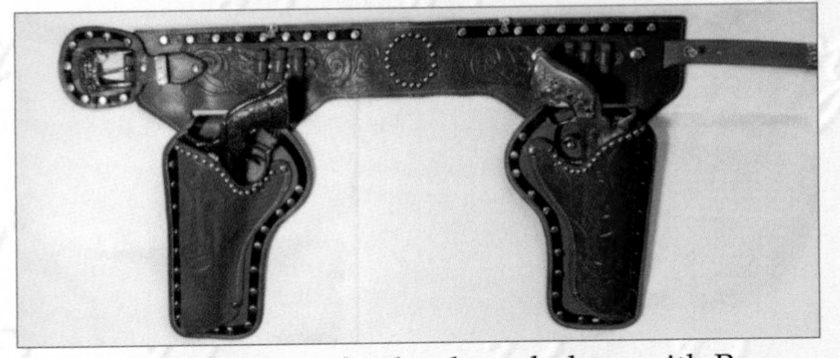

Roy Rogers brown leather large holster with Roy Rogers studs and tooled leather trimmed in black, 1955. (C.Q.)

Value $200-350 holster only

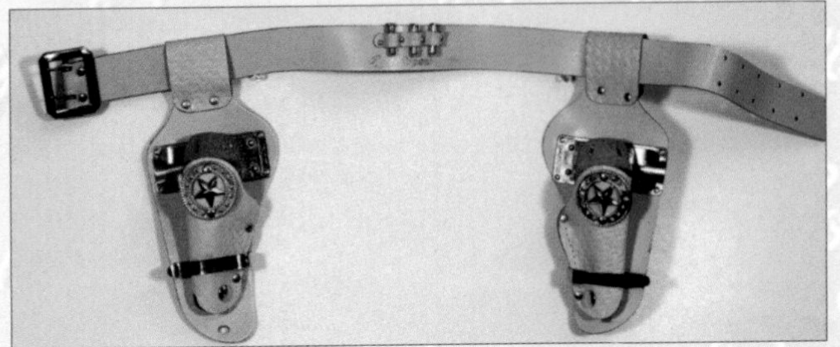

Classy "Flash Draw" holster, extremely rare! Tan leather with Roy's name under belt bullet loops, star conches on holster pockets. (C.Q.)

Value $700-1,100

Black and yellow leather holster with four Roy Rogers medallions, twin holster pocket buckles, eight bullet loops, and name in mid-section of belt. Perforated diamond-shaped cutouts, red jewels and numerous small studs. (G.S.)

Value $185-300 holster only

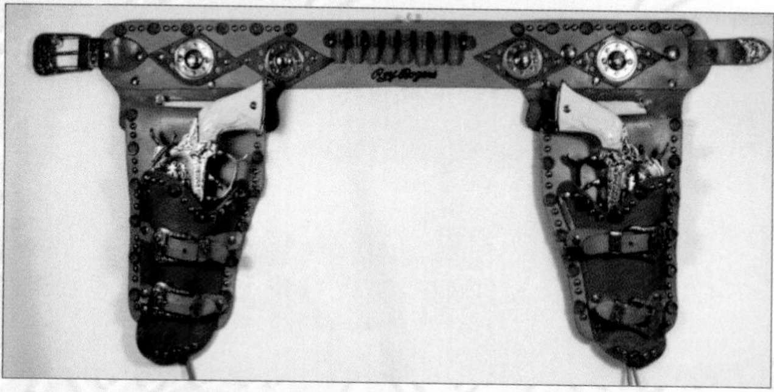

Brown and tan leather holster with four Roy Rogers medallions, twin buckles on holster pockets, right bullet loops, name on belt, cutout diamond-shaped leather overlays, many small studs and red plastic jewels. (C.Q.)

Value $185-300 holster only

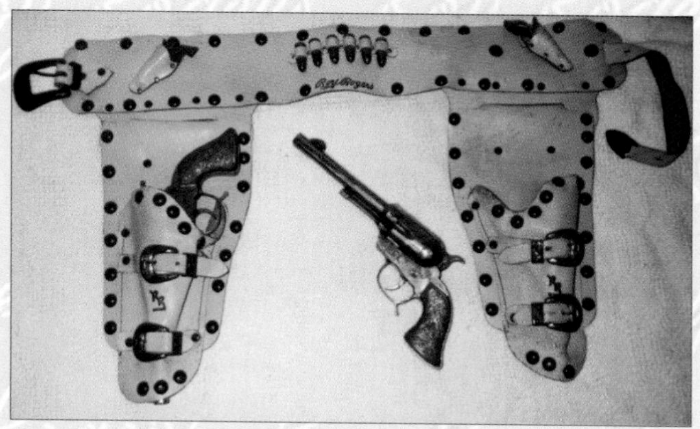

Unusual girl's large holster in white, gold-colored twin holster pocket buckles, large studs, and pair of miniature guns and holster pockets on belt. Roy's name under bullet loops. (G.S.)

Value $250-350 holster only

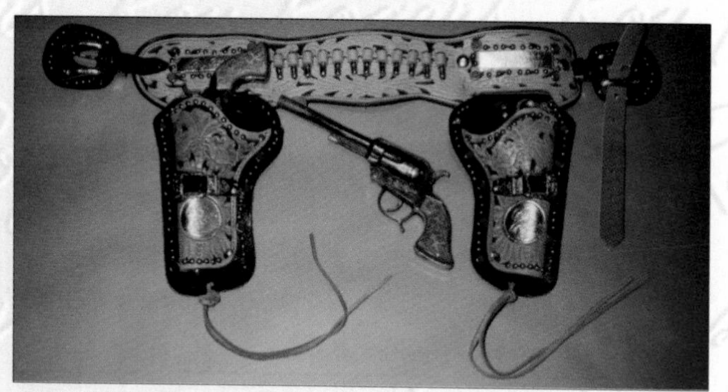

Classy large rare tooled tan and russet holster with silver foil overlay medallions, twin buckles, and holster pocket buckles. (G.S.)

Value $300-600 holster only

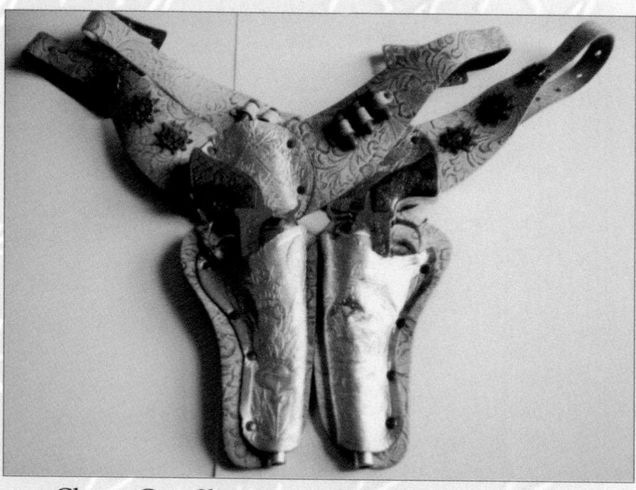

Classy Gun Slinger twin holster with gold foil overlaid on embossed leather with Roy's name on the center leather medallion of each holster. (M.M.)

Value $450-850 holster only

Classy large extremely rare "Flash Draw" circa 1958-1959, tooled brown leather with black trim, eight bullet loops in belt, and "Roy Rogers" on holster pockets. (G.S.)

Value $750-1,200 holster only

Brown holster with black trim, tooled leather with Roy Rogers studs, twelve bullet holders, and holster pocket buckles. (C.Q.)

Value $200-350

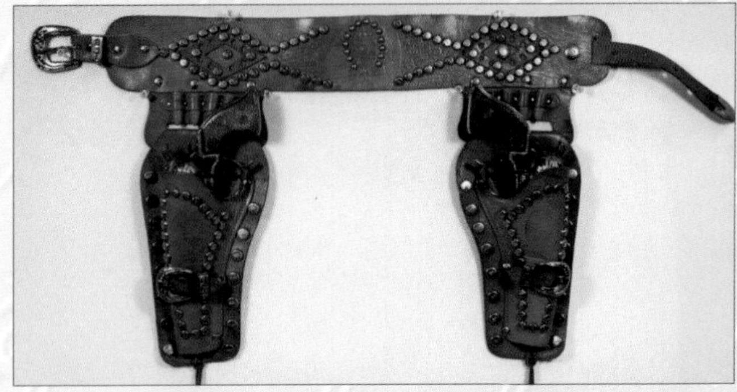

Halco large brown leather holster with many large and medium-sized Roy Rogers studs. Roy's image is embossed in studded horse shoe in middle of the 3-inch wide belt. (C.Q.)

Value $250-350 holster only

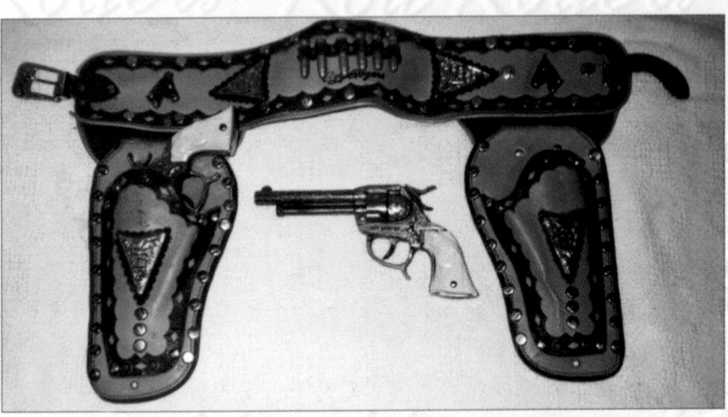

Brown and tan holster with Roy Rogers triangle medallions, scalloped leather trim, Roy Rogers studs and belt horse head overlaid leather. (G.S.)

Value $250-350 holster only

Small black single holster with "King of the Cowboys" imprinted with red jewels. (C.Q.)

Value $35-75 holster only

Single holster for small single shot cap gun. (C.Q.)

Value $25-45 holster only

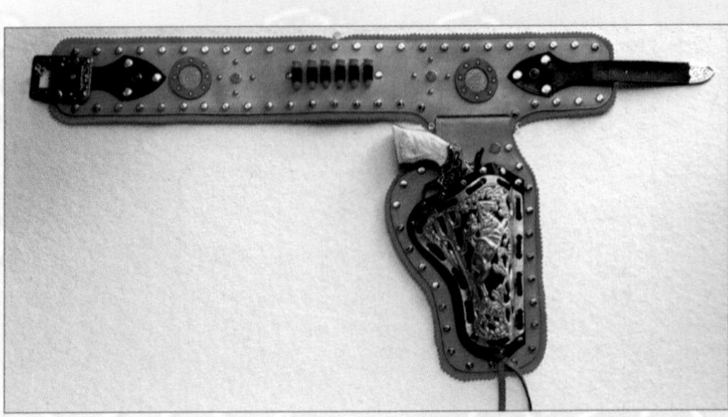

Esquire extremely rare and beautiful large fancy tan holster with red and black trim. Numerous Roy Rogers' studs and look at the elaborate metal holster pocket, complete with red leather tie-downs. (C.Q.)

Value $450-750 holster only

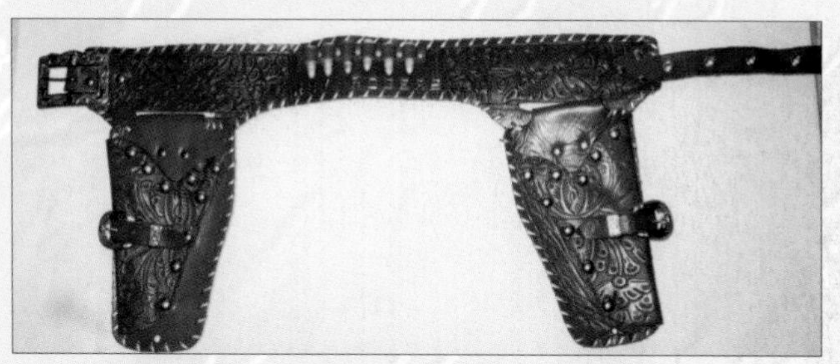

Large tooled brown leather holster with lacing on outer edge of belt and pockets. (C.Q.)

Value $250-350

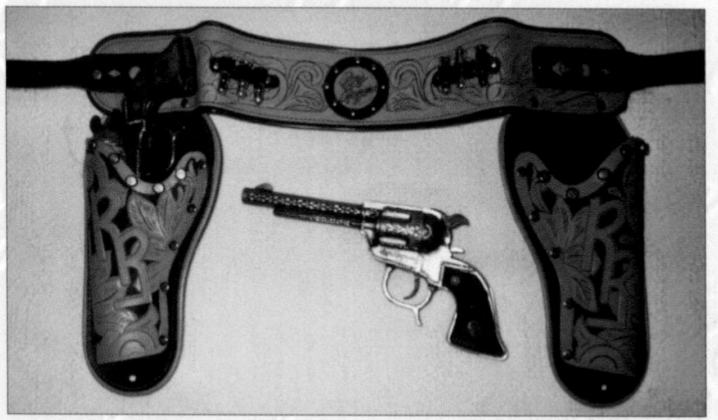

Classy tan and russet tooled holster with Roy Rogers studs and name in the center of the belt. (G.S.)

Value $225-325 holster only

Leslie Henry black and white rare holster with silver-colored leather arrows; a tan, black and white holster was also produced. Both have red plastic accent jewels, 1955. (B.W.)

Value $225-375 holster only

Tan and brown leather holster with Roy Rogers studs and cutouts in belt. (C.Q.)

Value $180-250 holster only

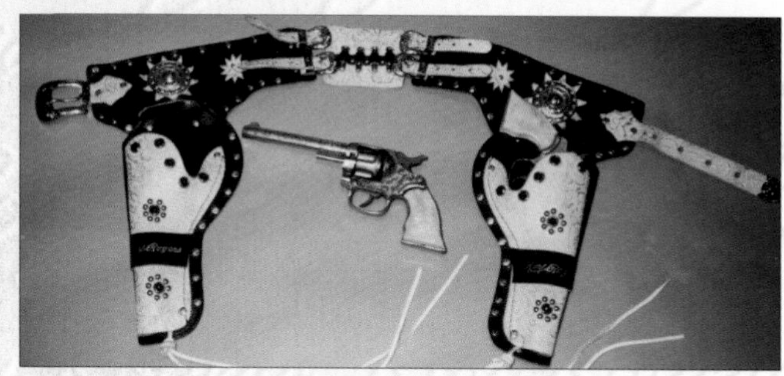

Large black and white fancy tooled leather holster with three-piece belt, star-type designs around large Roy Rogers conches, note four small buckles with adjustment straps on belt, Roy Rogers name on each holster pocket strap. (G.S.)

Value $250-350 holster only

Esquire tooled leather large holster, like the one Roy wore in the movies, very sought after, rig has a buckle over the pockets and holds 16 toy bullets. (C.Q.)

Value $200-350 holster only

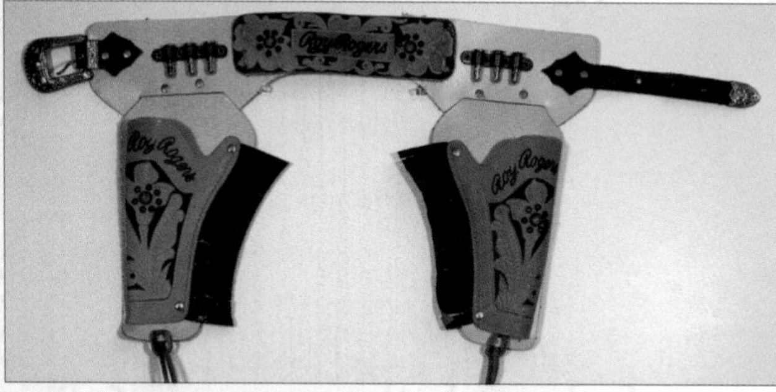

Tooled brown and tan holster with fringes 1955-1960. Roy's name with jewels and studs in the middle of the belt. (C.Q.)

Value $180-250

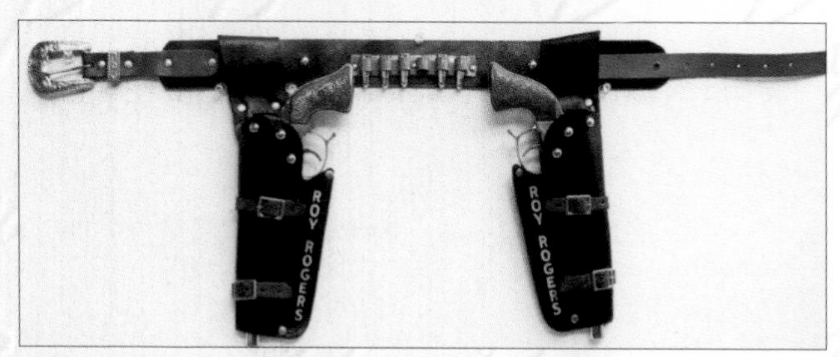

Plain black and brown holster with Roy's name and twin buckles on pockets. (C.Q.)

Value $100-175 holster only

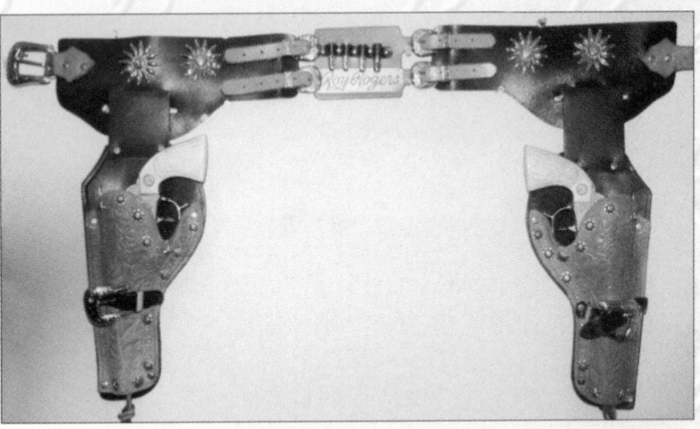

Very large tan and black leather holster with tooled pockets, four adjustable belts and buckles in this three-piece holster. Belt has star conches and "rosette" studs and buckle straps on pockets. (B.W.)

Value $250-350 holster only

Tan and black holster with tooling, Roy Roger's name on holster pockets, red jewels set in circles of studs. (C.Q.)

Value $180-280 holster only

Rare Keystone Brothers small brown holster with adjustable belt ties and Roy Rogers "King of the Cowboys" with large conches, 1950s. (G.S.)

Value $175-250 holster only

Classy leather "Flash Draw" embossed brown holster, rare. (G.S.)

Value $500-700 holster only

Girl's one gun white very fancy holster trimmed with numerous studs, red jewels, and overlay black and white conches. Very rare! Roy's name under bullet holders in metal letters and on holster pocket. (C.Q.)

Value $350-450 holster only

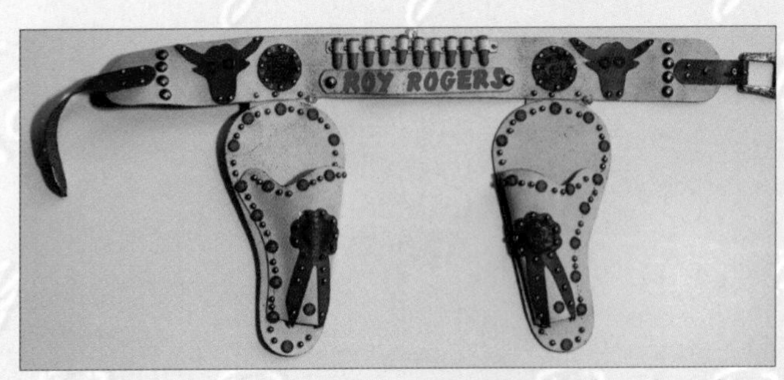

Rare girl's white small holster trimmed with red leather overlay conches, studs, and red jewels. (C.Q.)

Value $125-250

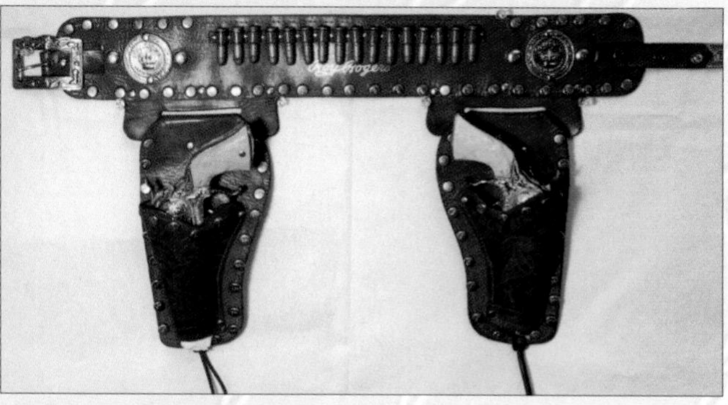

Large medium brown holster with Roy Rogers studs and large medallions on both sides of belt, bullet loops hold 16 bullets. (C.Q.)

Value $250-400 holster only

Large tooled holster with red jewels, black and tan with white fringes. (G.S.)

Value $225-235 holster only

Small red and tan single holster with Roy and Trigger's image silk-screened on holster pocket. (C.Q.)

Value $75-125 holster only

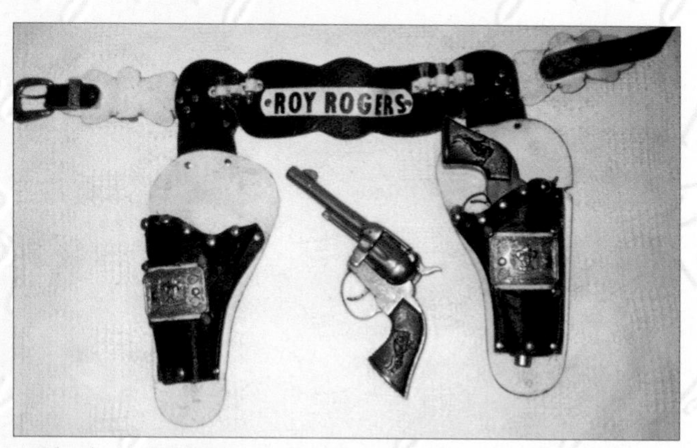

Black and white small holster with rectangular Roy Rogers medallions and red jewels on holster pockets. (G.S.)

Value $125-175 holster only

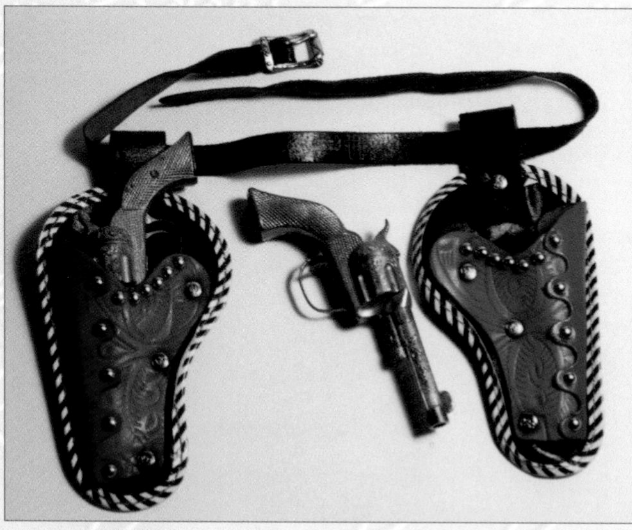

Very small Classy black and brown holster with white lacing on holster pockets made for single shot small cap guns. (C.Q.)

Value $75-125 holster only

Boxed Keystone set. (M.M.)

Value $900-1,200

Assortment of Classy Products Corp. various-sized Roy Rogers Holster sets. (G.S.)

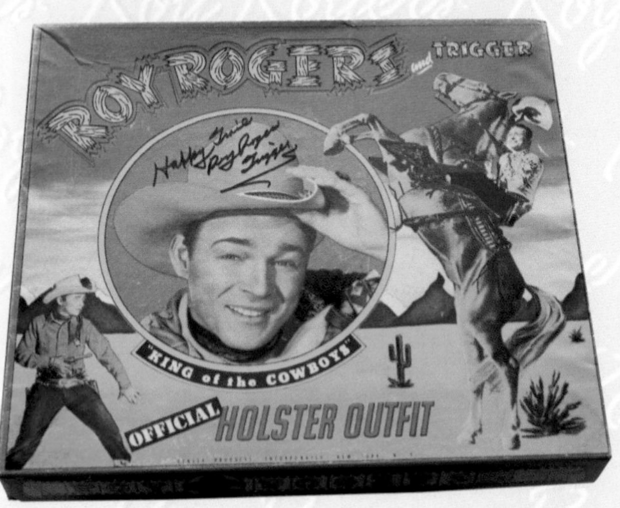

Early 1950s Classy Roy Rogers and Trigger 13 1/2″ x 12″ autographed holster outfit two-piece box. (Roy Rogers Museum)

Value $150-275 box only; add $100 for autograph

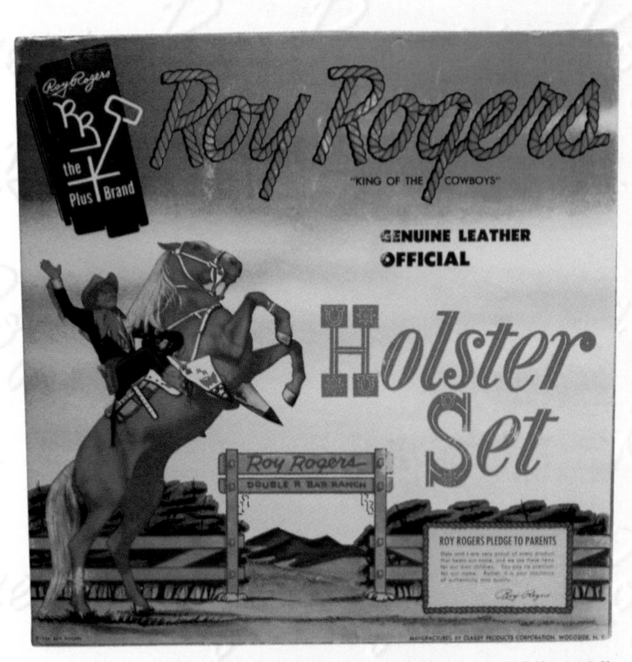

Late 1950s Classy Roy Rogers and Trigger 13″ x 12″ holster set box. (C.Q.)

Value $125-250 box only

Classy "Flash Draw" autographed 13 1/2″ x 12″ holster outfit box (C.Q.)

Value $175-350 with autograph

Value $125-275 box only

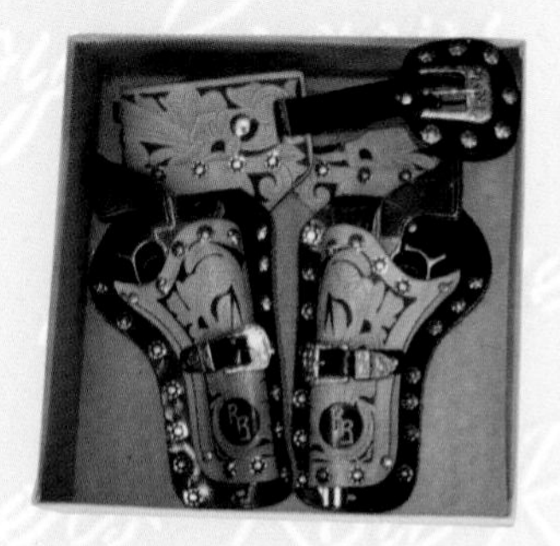

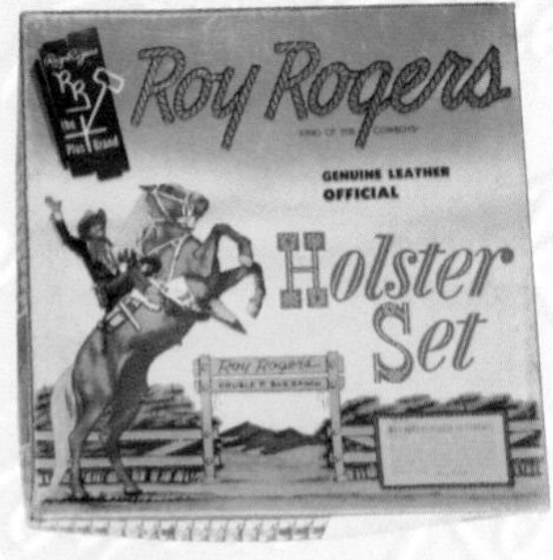

Classy black and tan holster with George Schmidt cap guns set. (B.W.)
Value $1,250-1,650

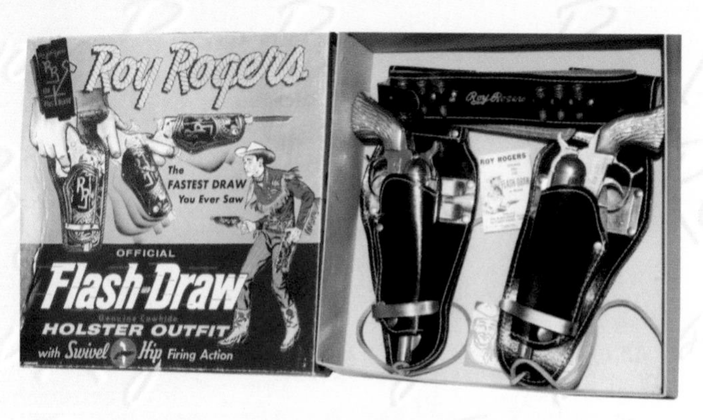

Classy "Flash-Draw" black holster outfit with instructions, etc., very rare. (B.W.)
Value $1,400-1,850

Daisey Mfg. Co., Ltd. of Canada display box with holster and die-cast Kilgore cap gun—note turquoise holster trim, a very unique set. (B.W.)
Value $250-450

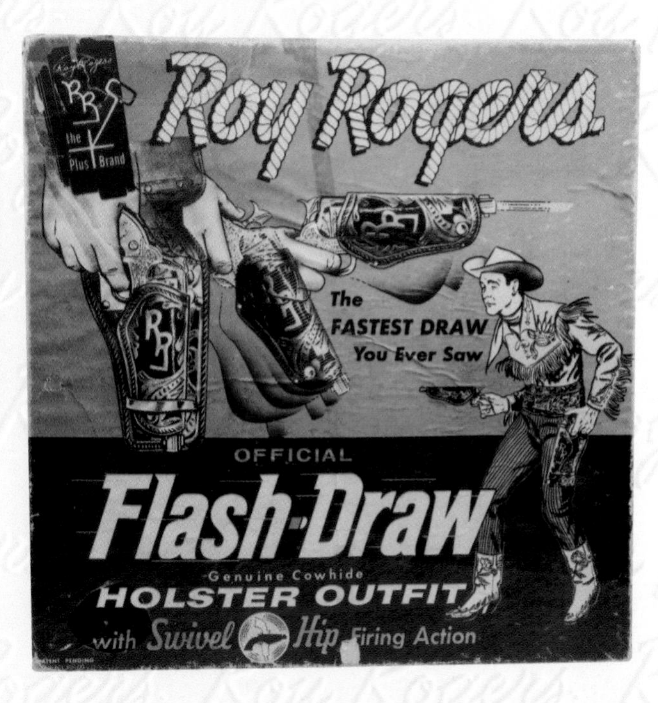

Classy "Flash-Draw" 13 1/2" x 12" holster box. (C.Q.)
Value $125-275

Fancy studded and tooled leather holster with "Cowboy King" die-cast cap guns and box. (L.Z.)
Value $850-1,200 complete set

Classy holster and gun set with two-piece box. (B.W.)
Value $450-650

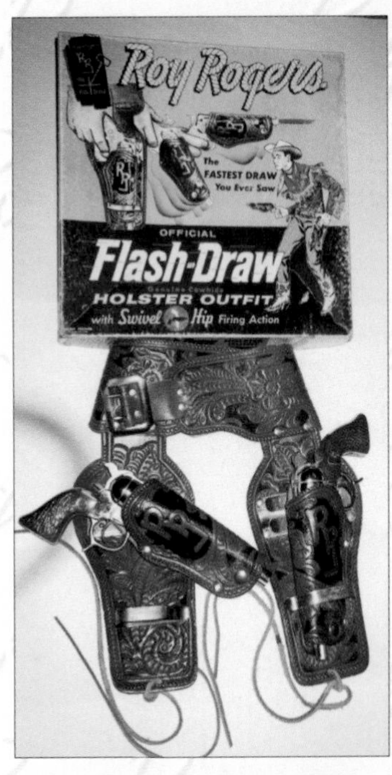

Classy "Flash Draw" extremely rare holster and gun outfit with very fancy tooling on the holster. (G.S.)
Value $1,400-1,850 complete set

Early 1950s Classy 9" x 13" rare Roy Rogers and Trigger holster outfit box (R.L)
Value $150-275

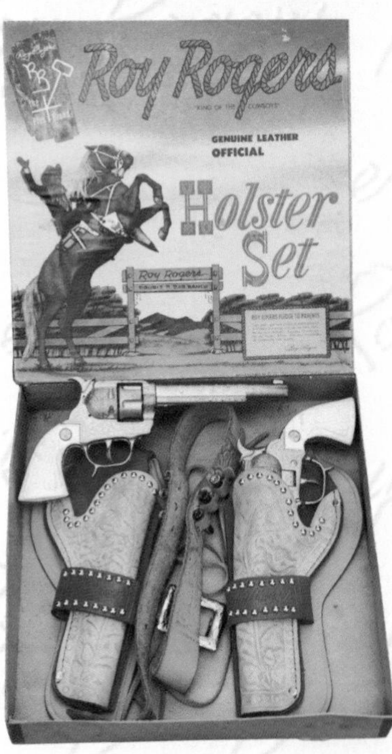

Classy large tooled leather holster with 12" Classy cap guns in original box. (B.W.)
Value $1,200-1,800

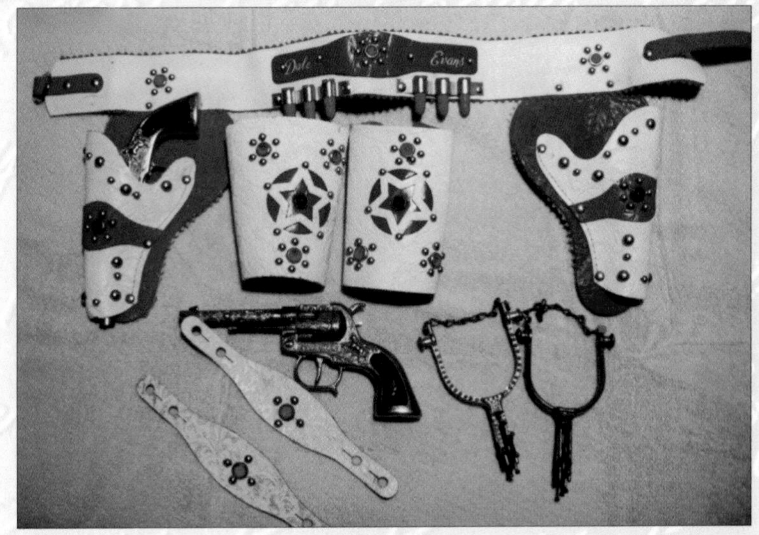

Dale Evans extremely rare Buzz Henry complete set including holster, spurs, cuffs, and guns. (B.W.)
Value $1,200-1,800 for set

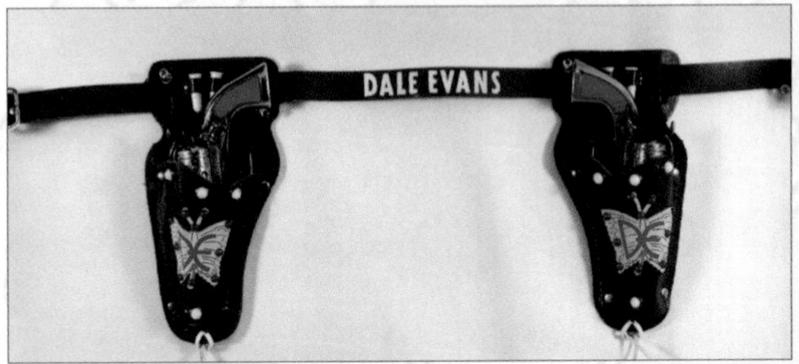

Dale Evans Classy very rare small black holster with red and white jewels and butterfly overlay with red letters on holster pockets. (C.Q.)
Value $550-850

Dale Evans Classy red-and white-leather medium size leather holster with red jewels set in conches and white fringe on holster pockets. (B.W.)
Value $1,200-1,800 for set

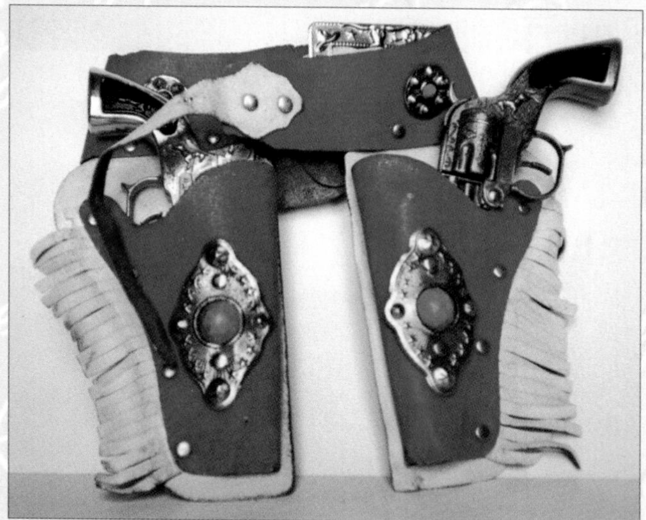

Dale Evans red-and white-ringed holster with Buzz Henry gold-plated guns. (M.M.)
Value $1,200-1,500

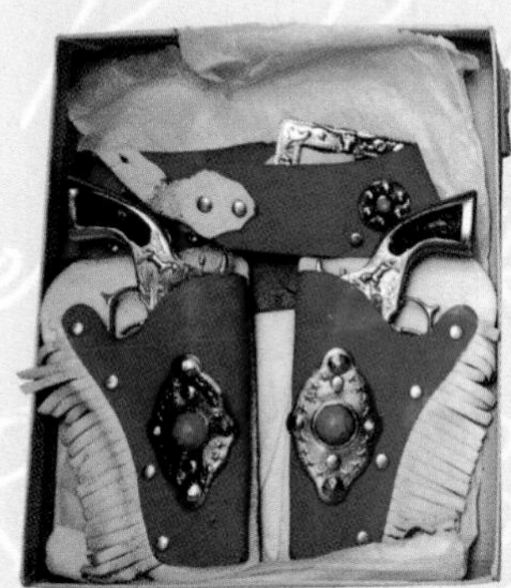

Dale Evans extremely rare holster and Buzz Henry cap guns in original early 1950s two-piece box. (M.M.)
Value $1,500-2,000 complete set

Dale Evans very rare two-piece holster set box. (C.Q.)
Value $250-400 box only

Dale Evans very rare Classy box with very rare red and white holster and Buzz Henry cap gun. (G.S.)
Value $1,000-1,800 complete set
Value $900-1,200 George Schmidt cap guns with red jewels

Dale Evans very rare small red leather holster with white leather overlay trim, and Dale Evans Buzz Henry cap gun. (C.Q.)
Value $450-750 complete set

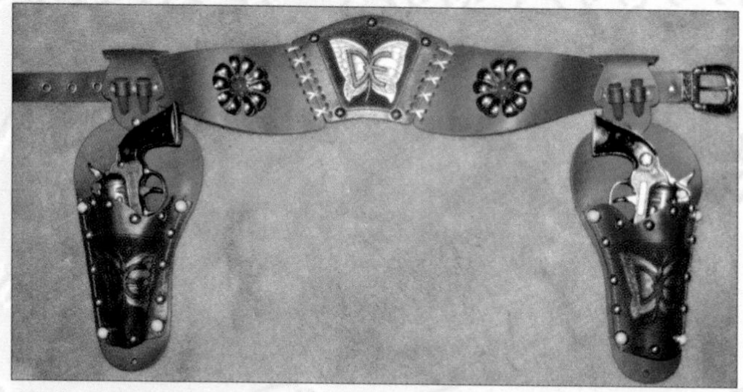

Dale Evans Classy very rare medium-size 1960s green leather holster with white Dale Evans butterfly embossed leather symbols. (B.W.)
Value $600-950 holster only

Roy Rogers jack knife retail counter display. (C.Q.)
Value $150-200 display only

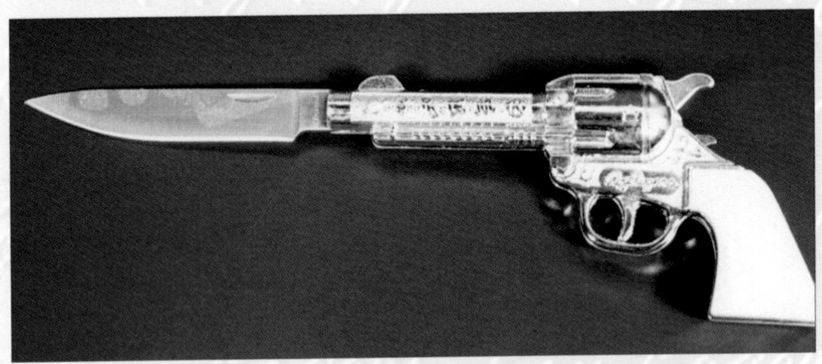

Roy Rogers 8" gun knife with blade that folds into the gun, very rare. (C.Q.)
Value $150-225

Red-handled Roy Rogers and Trigger jack knife with compass in sheath. (C.Q.)
Value $75-125

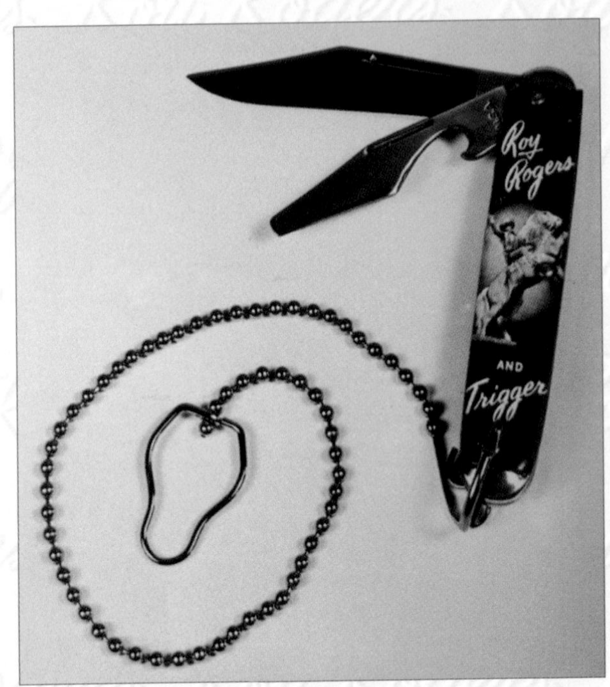

Black plastic handled jack knife on beaded key chain with Roy Rogers name and image on the handle. (C.Q.)
Value $60-125

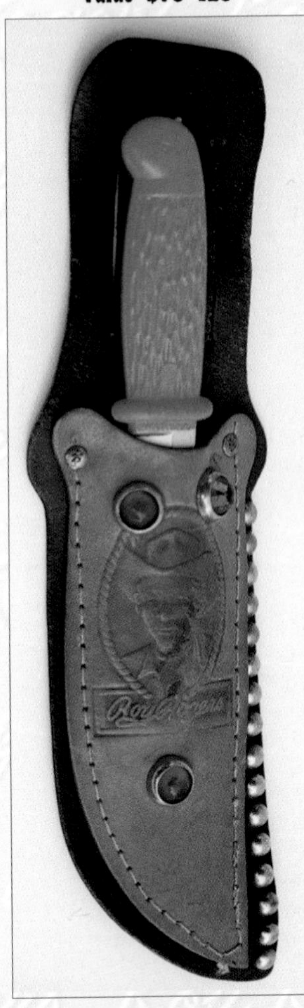

Roy Rogers rubber knife with red-jeweled sheath, 7 1/2" long. (C.Q.)
Value $45-90

ROY ROGERS MARX COMPANY RIFLES

Marx silver plastic 25″ Deluxe Winchester carbine with octagon barrel, with metal cap-firing components and metal cocking lever. Roy's name etched in gold on the stock, 1950s. (C.Q.)

Value $125-250

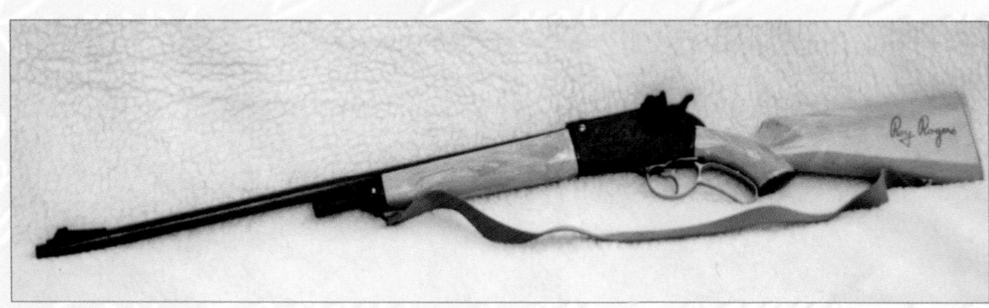

Marx Deluxe Winchester Model 71 with gold-etched name on stock and sling, 30″, plastic stock and forearm. (C.Q.)

Value $150-300

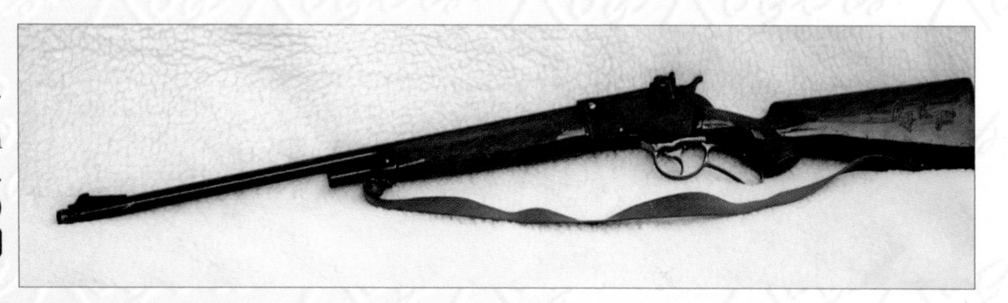

Marx Deluxe Winchester 30″ dark brown plastic rifle with metal cap firing components. (C.Q.)

Value $150-300

Marx box for Deluxe Winchester silver plastic carbine. (G.S.)

Value $60-120

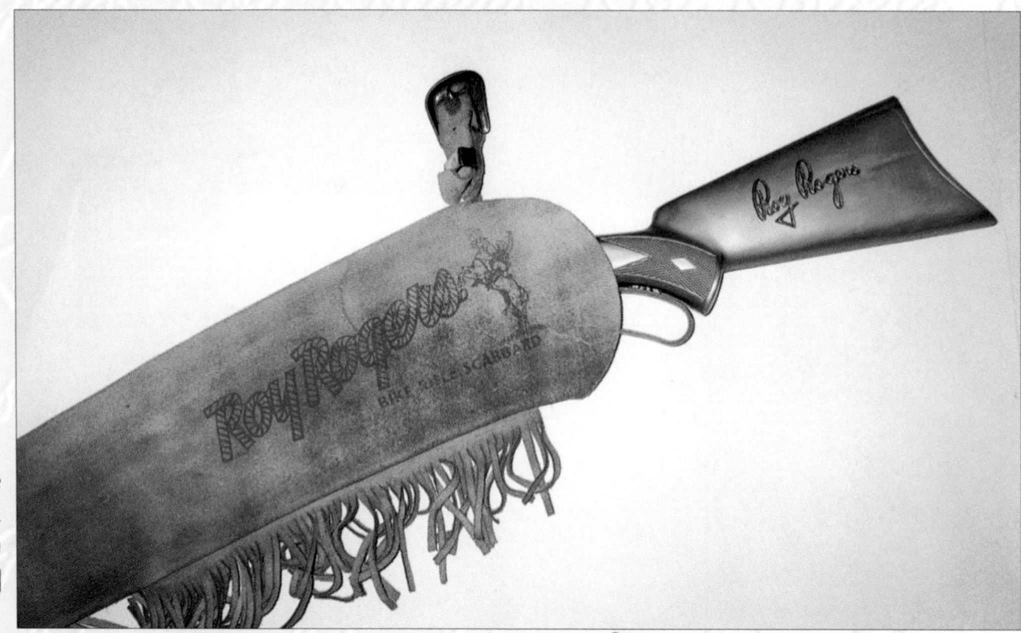

Bike rifle case made of leather to hold carbine. (G.S.)

Value $75-150

CHAPTER 6 CARDS

More than 20 companies, including Republic Pictures and General Foods Post Cereals, printed all types of photo and trading cards with images of Roy Rogers, Dale Evans, and Trigger. These cards were made not only in the United States, but in other countries as well.

Republic Pictures started the Roy Rogers card craze to promote their western hero star in the early-to mid-1940s. Other companies soon joined the bandwagon and produced their own cards. Some were sold in sets as trading cards and others were offered in a pack of gum, a cereal box, or the penny arcade machine. Kane Products of England put collectible images of Roy on their boxes of candy.

Many different sets of foreign and domestic cards exist. Most of the Roy Rogers photos used on trade and photo cards were taken from Republic Pictures publicity shots of the late 1940s to the mid-1950s. Complete numbered card sets generally have more value per card than loose single cards.

ARCADE CARDS

Arcade exhibit photo cards. (C.Q.)
Value $15-25 each

Roy Rogers penny arcade vending machine supplied by The Exhibit Supply Company in the 1950s. Put one or two pennies in the coin-holder, push it in, and a card would pop out of a slot. Arcade exhibit cards were usually monochrome black and white publicity photos of movie stars printed on cardboard. Extremely rare. (G.S.)

Arcade exhibit photo cards 3 5/16" x 5 3/8". (G.S.)
Value $15-25 each

Roy Rogers arcade exhibit photo cards. (C.Q.)
Value $15-25 each

Early vintage sepia tone post cards of Dale Evans, Roy Rogers, and Gabby Hayes, rare. (C.Q.)

Value $25-45 each

3-D picture card from inside box of Post Sugar Crisp Cereal.
There were four different ones in a set 2 1/2" x 4 1/4". (C.Q.)

Value $10-15; complete set with viewer value $55-80

Kane Products, Ltd., England, 1950s candy box with photos of Dale Evans, rare. (C.Q. and M.M.)

Value $30-40

Assorted collection of late 1940s and 1950s premium cards ranging in size of 3 1/2" x 4 1/2" in length. (R.L.)

Value $5-15 each

Close-up view of one-penny arcade exhibit card vending machine. (G.S.)

Roy Rogers and Trigger post card 1950s. (C.Q.)

Value $20-25

A.B.C. MINORS
PICTURE CARDS
Set 1. Series of 10
FILM STARS
No. 1
Roy Rogers
King of the Cowboys, and Trigger, the smartest horse in the movies, take Bullet the dog for a ride.

WEAR YOUR MINORS BADGE FOR A PICTURE CARD
•
COLLECT THE SERIES FOR A SPECIAL BADGE

ABC Minors' picture card, front and back views. (C.Q.)

Value $15-20

Roy Rogers and Trigger movie photo trading card. (C.Q.)

Value $12-15

HOLLYWOOD
DALE EVANS

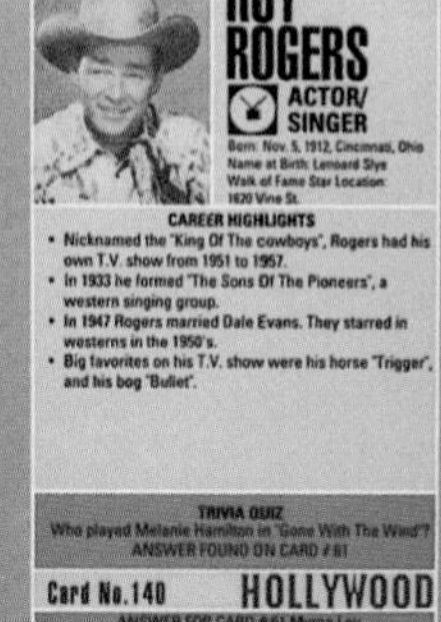

Trading cards by Hollywood front and back. (C.Q.)

Value $18-25 each

CHAPTER 7

CHILDREN'S STORYBOOKS

Whitman Publishing Company and Simon and Schuster were two of the largest companies that published Roy Rogers and Dale Evans children's books from the 1940s through the 1950s. Whitman Publishing Company published Roy and Dale hard back novels, Better Little Books, Cozy Corner Books, Tell-a-Tale Books, and a variety of children's activity books. The first Roy Rogers storybooks published by Whitman Publishing Company were Better Little Books between 1942 and 1950. *Robinhood of the Range* was the first book of the series.

BETTER LITTLE BOOKS

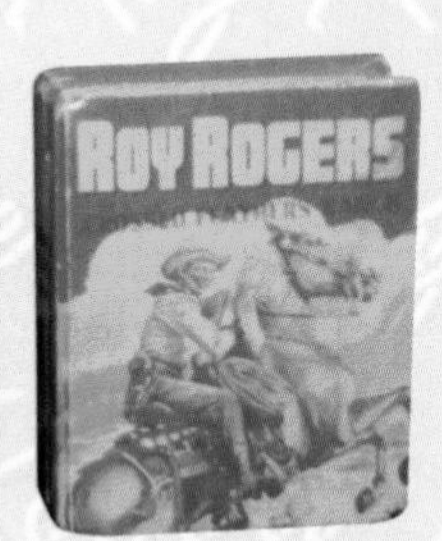

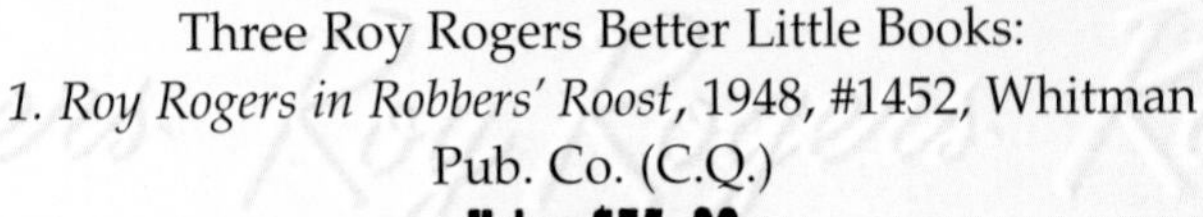

Three Roy Rogers Better Little Books:
1. *Roy Rogers in Robbers' Roost*, 1948, #1452, Whitman Pub. Co. (C.Q.)
Value $35-60

2. *Roy Rogers and the Snowbound Outlaw*, 1948, Whitman Pub. Co. Tall Better Little Book, 3 3/16" x 5 7/16". (C.Q.)
Value $35-60

3. *Roy Rogers and the Mystery of Howling Mesa*, 1948, Whitman Pub. Co. (C.Q.)
Value $35-60

Three Roy Rogers Better Little Books:
1. *Roy Rogers and the Dwarf Cattle Ranch*, 1947, Whitman Pub. Co. (C.Q.)
Value $35-60

2. *Roy Rogers, Range Detective*, 1950, Tall Better Little Book, Whitman Pub. Co. (C.Q.)
Value $35-60

3. *Roy Rogers at the Crossed Feathers Ranch*, 1945, Whitman Pub. Co. (C.Q.)
Value $35-60

Original illustrations by Edwin Hess for Whitman Pub. Co. Better Little Book, *Roy Rogers at the Crossed Feathers Ranch*. Extremely rare. (G.S.)
Value unknown.

Not shown: *Roy Rogers and the Deadly Treasure*, 1947, Better Little Book, Whitman Pub. Co.
Value $35-60

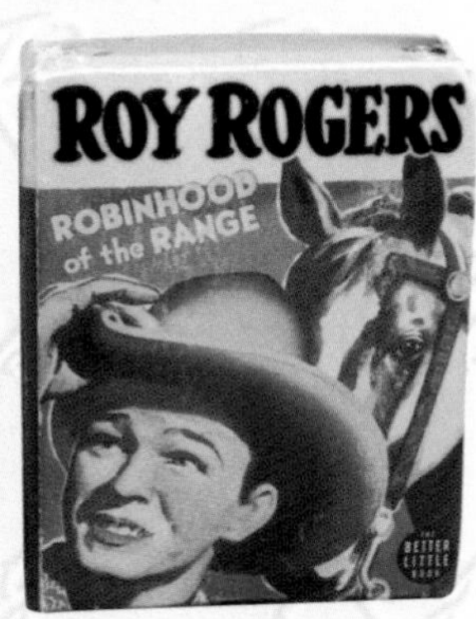

Roy Rogers, King of the Cowboys, 1943, and *Robinhood of the Range*, 1942, Better Little Book, Whitman Pub. Co. (C.Q.)
Value $35-65 each

HARDBACK NOVELS

Roy Rogers and the Gopher Creek Gunman. (C.Q.)
Value $25-50

Roy Rogers and the Ghost of Mystery Rancho, 1950. Whitman Pub. Co. (R.L.)
Value $20-40
(add $10 for dust jacket)

Roy Rogers and Dale Evans in River of Peril, 1957. Whitman Pub. Co. (C.Q.)
Value $25-45

Roy Rogers, King of the Cowboys, 1956. Whitman Pub. Co. (R.L.)
Value $20-50

Roy Rogers and The Enchanted Canyon, 1954. Whitman Pub. Co. (R.L.)
Value $20-50

Roy Rogers and the Brasada Bandits, 1955. Whitman Pub. Co. (R.L.)
Value $20-50

Roy Rogers and the Rimrod Renegades, 1952. Whitman Pub. Co. (R.L.)
Value $20-40
(add $10 for dust jacket)

Roy Rogers and the Outlaws of Sundown Valley, 1950. Whitman Pub. Co. (C.Q.)
Value $20-40
(add $10 for dust jacket)

Other hard back novels 5 1/4″ x 8″ published by Whitman Publishing Company:

Roy Rogers and the Raider of Sawtooth Ridge, 1946.
Value $20–50

Roy Rogers and the Trail of Zeros, 1954.
Value $20–50

Dale Evans and Danger in Crooked Canyon, 1958.
Value $20–50

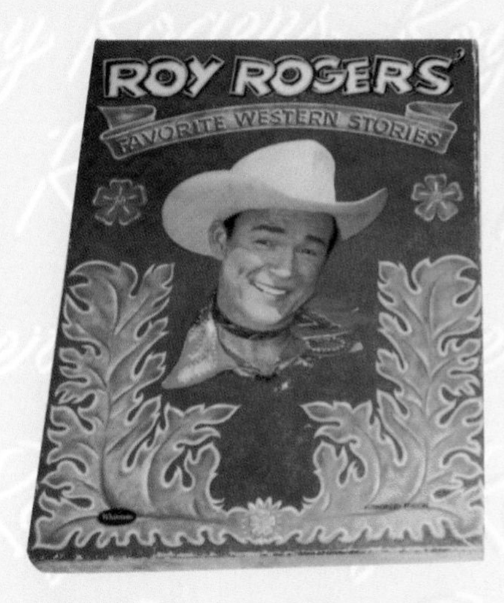

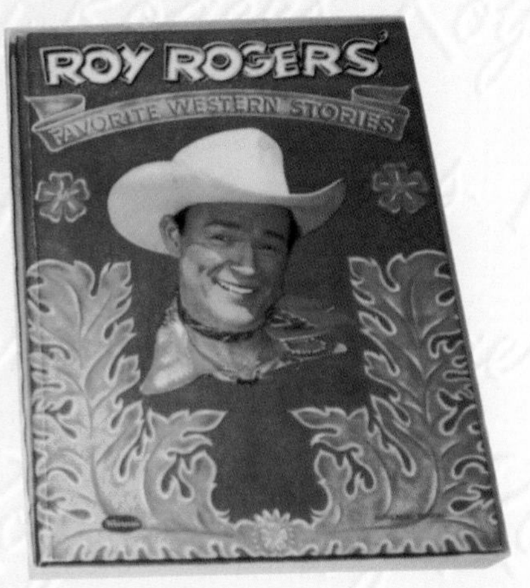

Roy Rogers' Favorite Western Stories, with 7" x 10" box, over-sized book. Whitman Publishing Company. (M.M.)
Value $45-65

Dale Evans' Favorite Bible Stories, book size 8 1/2" x 11". Whitman Publishing Co. (R.L.)
Value $15-30

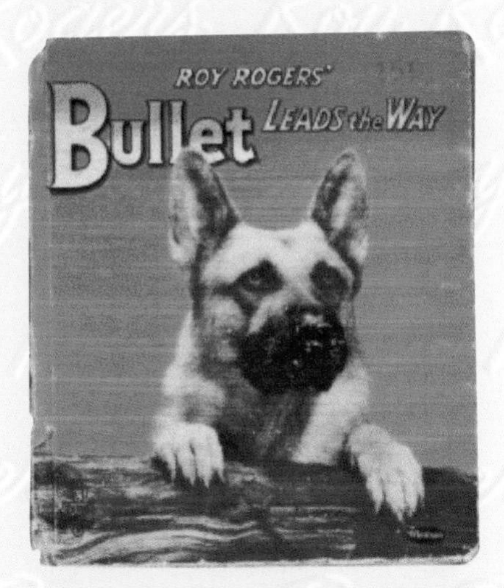

Roy Rogers' Bullet Leads the Way, 1953, Tell-a-Tale Book, 5 1/2" x 6 1/2", Whitman Publishing Company. (R.L.)
Value $15-30

Dale Evans and Buttermilk, 1956, Tell-a-Tale Book, 5 1/2" x 6 1/2", Whitman Publishing Company. (C.Q.)
Value $15-30

Other Tell-a-Tale Books published by Whitman Publishing Company:

Roy Rogers in Surprise for Donnie, 1954.
Value $15–30
Roy Rogers at the Lane Ranch, 1950.
Value $15–30
Roy Rogers and the Sure 'Nough Cowpoke, 1952.
Value $15–30

SIMON AND SCHUSTER BIG GOLDEN BOOKS AND LITTLE GOLDEN BOOKS BY SANDPIPER PRESS AND ARTISTS AND WRITERS GUILD

Roy Rogers, King of the Cowboys, 1953. Simon and Schuster Big Golden Book. (C.Q.)
Value $30-75

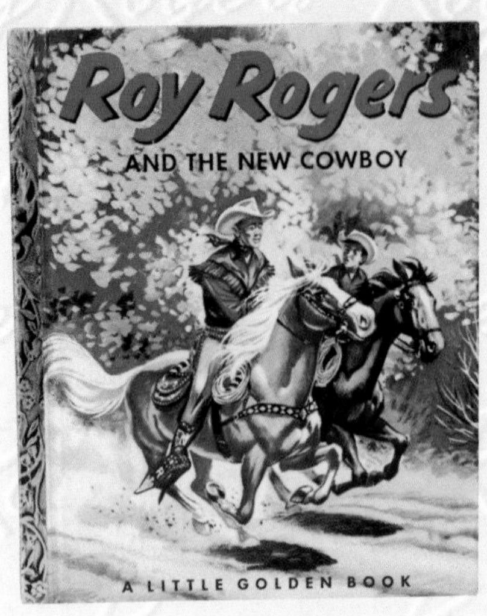

Roy Rogers and the New Cowboy, 1953. Simon and Schuster Little Golden Book. (C.Q.)
Value $25-50

Roy Rogers and Cowboy Toby, 1954. Simon and Schuster Little Golden Book. (C.Q.)
Value $25-50

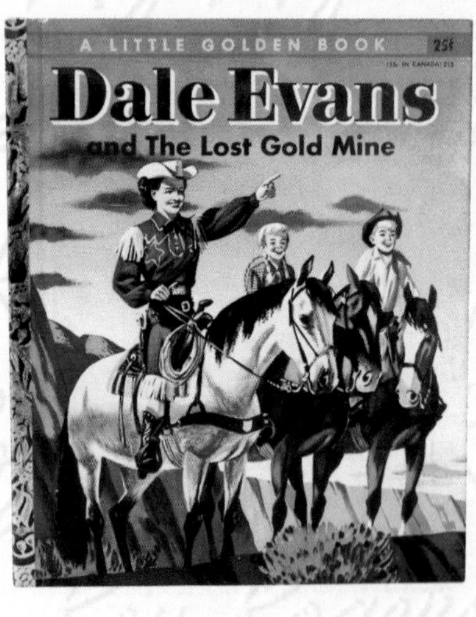

Dale Evans and The Lost Gold Mine, 1954. Simon and Schuster Little Golden Book. (C.Q.)
Value $25-50

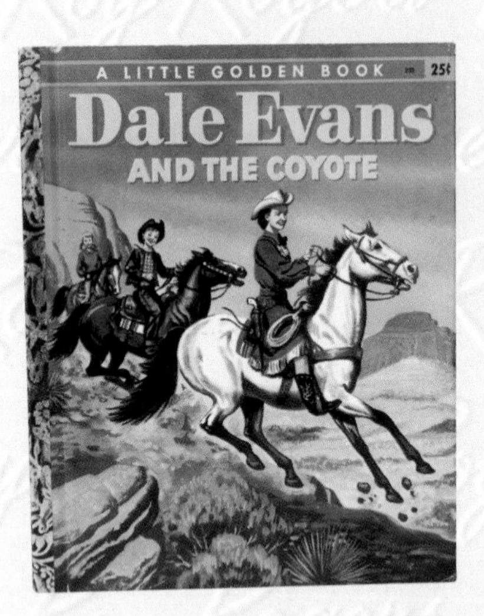

Dale Evans and The Coyote, 1956. Simon and Schuster Little Golden Book. (C.Q.)
Value $25-50

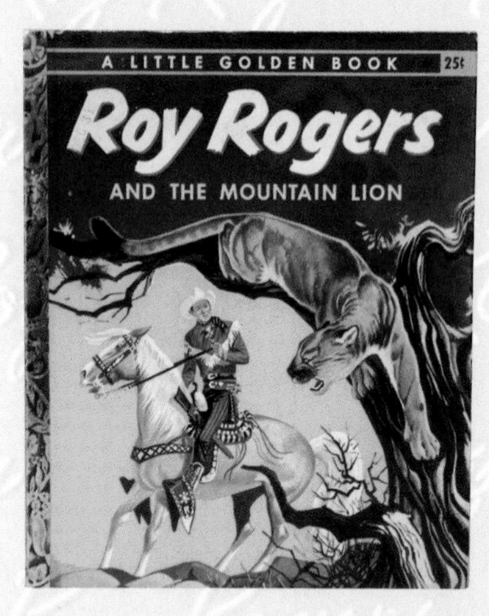

Roy Rogers and the Mountain Lion, 1955. Simon and Schuster Little Golden Book. (C.Q.)
Value $25-50

Other Golden Books published by Simon and Schuster:

Roy Rogers and the Indian Sign, Little Golden Book.
Value $25–50
Prayer Book for Children, Big Golden Book, 1956.
Value $30–75

CHAPTER 8

COLORING AND ACTIVITY BOOKS

Whitman Publishing Company issued a great variety of children's coloring and activity books that featured Roy Rogers, Dale Evans, Roy Rogers, Jr., Gabby Hayes, Pat Brady, Trigger, and Bullet in the late 1940s through the 1950s. It is very difficult to locate any of these books in an unused mint condition, as these books were created for hard use. Condition is of prime importance in valuing these types of books. Cutout and other activity books are considered more rare than color and paint books.

Estimated values for books shown in this chapter are for paint and color books that have few to no pictures painted or colored. Values listed for cutout and activity books shown are for books that have few or no missing pieces.

COLORING AND PAINT BOOKS

(Note: Book dimensions range from 8" x 12" to 11" x 15")

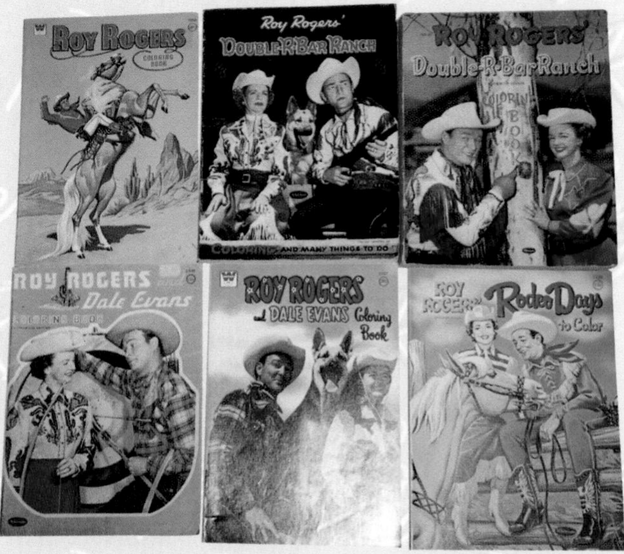

Collection of Roy Rogers and Dale Evans coloring books, 1950s. Whitman Publishing Company. (G.S.)

Value $35-75 each

Collection of Roy Rogers and Dale Evans coloring books, 1950s. Whitman Publishing Company. (G.S.)

Value $35-75 each

Dale Evans Coloring Book, 1950s, 11″ x 13 1/2″ by Whitman Publishing Company. (C.Q.)

Value $45-90

Roy Rogers and Dale Evans Coloring Book, 8″ x 11″ by Whitman Publishing Company. (C.Q.)

Value $35-75

Roy Rogers with Dale Evans, Trigger and Bullet Rodeo Sticker Fun Book, 1950s, Whitman Publishing Company. (G.S.)

Value $45-90

Roy Rogers' Trigger and Bullet Coloring Book, 1956, 8 1/2″ x 11″ by Whitman Publishing Company. (C.Q.)

Value $40-80

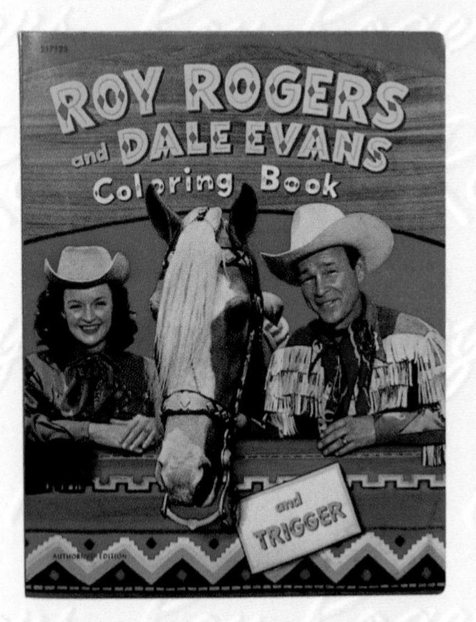

Roy Rogers, Dale Evans, and Trigger Coloring Book, 1950s, by Whitman Publishing Company. (C.Q.)

Value $35-75

Roy Rogers and Dale Evans Coloring Book. (C.Q.)

Value $35-75

Roy Rogers and Bullet Coloring Book, 8 1/2″ x 11″ 1953 by Whitman Publishing Company. (C.Q.)
Value $35-75

Roy Rogers and Dale Evans Coloring Book, 11″ x 15″ by Whitman Publishing Company. (C.Q.)
Value $45-90

Roy Rogers' Pal Pat Brady Coloring Book, 8 1/4″ x 10 1/2″ by Whitman Publishing Company. Rare. (C.Q.)
Value $40-80

Roy Rogers' Rodeo Days to Color, 8″ x 10 3/4″ by Whitman Publishing Company. (D.T.)
Value $35-75

Gabby Hayes Coloring Book, by Whitman Publishing Company. Rare. (G.S.)
Value $40-80

Gabby Hayes Magic Dial Funny Coloring Book, by Lowe Company. Rare. (M.M.)
Value $40-80

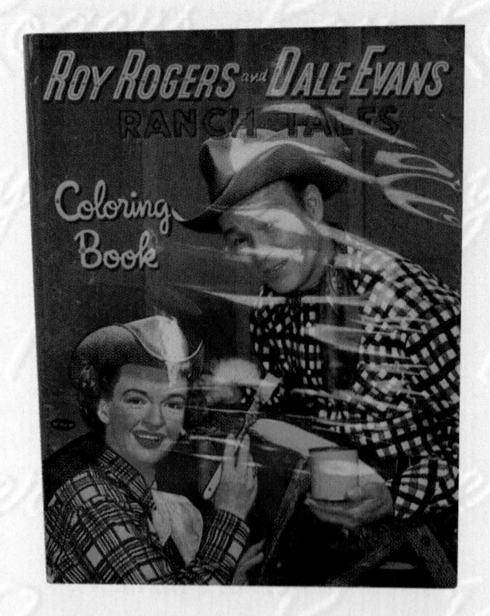

Roy Rogers, Dale Evans, and Trigger coloring books, 1950s, 8 1/2" x 11" by Whitman Publishing Company. (D.T.)
Value $35-75 each

A great collection of Roy Rogers and Dale Evans cutout doll and activity books, 1950s, Whitman Publishing Company. (G.S.)
Value $45-90 each

Roy Rogers Paint Book, 1948, Whitman Publishing Company. (G.S.)
Value $35-85

Collection of four Roy Rogers, Dale Evans, and Trigger coloring books, 1950s, by Whitman Publishing Company. (G.S.)

Value $35-75 each

Four Whitman Publishing Company cutout books:

1. *Roy Rogers and Dale Evans Cut-Out Dolls*, front cover folds out. (G.S.)
2. *Roy Rogers, Dale Evans, and Dusty Cut-Out Book*. (G.S.)
3. *Roy Rogers and Dale Evans Cut-Out Book*. (G.S.)
4. *Roy Rogers and Dale Evans Cut-Out Dolls*. (G.S.)

Value $45-90 each

Roy Rogers and Dale Evans Paint Book, 1950s, Whitman Publishing Company. (G.S.)

Value $35-85

Roy Rogers Paint Book, 1950s, Whitman Publishing Company. (G.S.)

Value $35-85

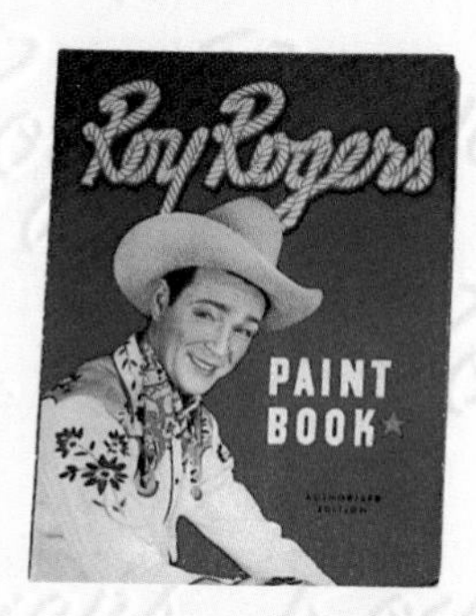

Roy Rogers Paint Book, 1948, Whitman Publishing Company. (G.S.)

Value $35-85

Roy Rogers and Dale Evans inside pages with figures of Roy and Dale with cut-out clothes. (G.S.)

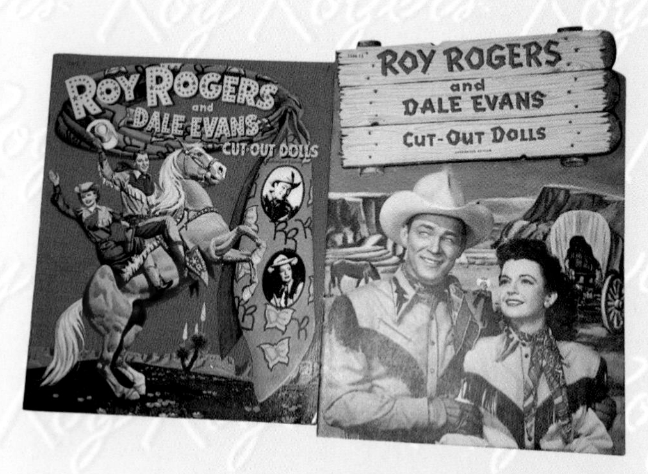

1. *Roy Rogers and Dale Evans Cut-Out Dolls Book*, 1950s, Whitman Publishing Company. (G.S.)

Value $45-90

2. *Roy Rogers and Dale Evans Cut-Out Dolls Book*, 1950s, Whitman Publishing Company. (G.S.)

Value $45-90

1. *Roy Rogers and Dale Evans Punch-Out Book*, 10" x 15" very rare, Whitman Publishing Company. (G.S.)

Value $75-150

2. *Roy Rogers and Dale Evans Punch-Out Book*, 10" x 15," 1952, Whitman Publishing Company. (G.S.)

Value $75-150

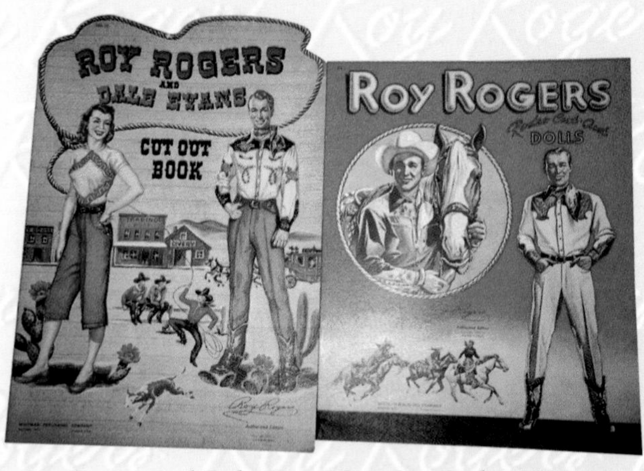

1. *Roy Rogers and Dale Evans Cut-Out Book*, 1950, Whitman Publishing Company. (G.S.)

Value $45-90

2. *Roy Rogers Rodeo Cut-Out Dolls Book*, 1953, Whitman Publishing Company. (G.S.)

Value $45-90

CHAPTER 9

Comic Books

Dell Publishing Company produced thirteen four color Roy Rogers comic books between 1944 and 1947. These comics are numbered 38, 63, 86, 95, 109, 117, 124, 137, 144, 153, 160, 166, and 177. Originally priced at 10 cents, they are now valued from $45 for issue #177 in C-8 condition to $2,400 for the first issue #38 in mint condition.

From 1948 to 1961, Dell Publishing Company issued 145 Roy Rogers comics that were produced by the Western Printing and Lithographing Company. Values for these comics range from $10 to $325 in C-8 to mint condition, with issue #1 being the most valuable and worth upwards of $800. The later issues have less value than the earlier issues.

Dale Evans comics were issued by Dell Publishing Company in 2 series during the 1950s. The first series had 24 issues, and the second series contained 22 issues. Both series are valued from $8 to $200 in C-8 to mint condition.

March of Comics issued 25 Roy Rogers comics from 1948 to 1963. These comic books featured store advertisements on the back cover and were given to customers of the store advertised. Sears, Roebuck & Company gave away thousands of these free Roy Rogers March of Comics.

Front and back covers of Roy Rogers March of Comics, 1950s. Photograph of Roy with gun and saddle on front cover. Advertisement by Sears, Roebuck & Co. for Roy Rogers and Dale Evans wrist watches on back cover. (G.S.)

Value $35-60

ROY ROGERS COMICS, DELL

#95 February 1946. (G.S.)
Value $275-400

#4 April 1948. (R.L.)
Value $125-175

#14 February 1949. (R.L.)
Value $60-85

#28 April 1950. (R.L.)
Value $50-75

#34 October 1950. (R.L.)
Value $40-55

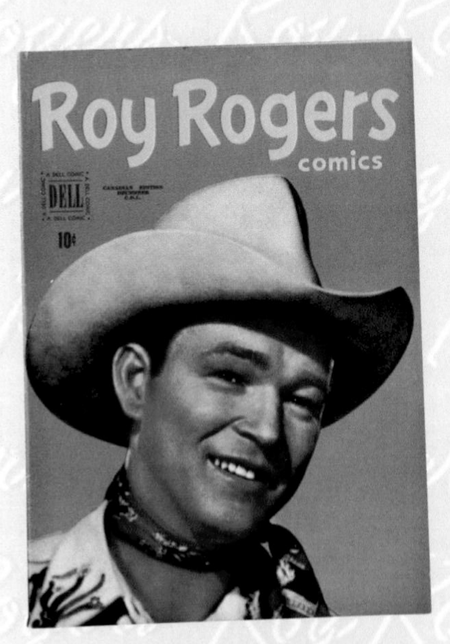

#35 November 1950. (R.L.)
Value $40-55

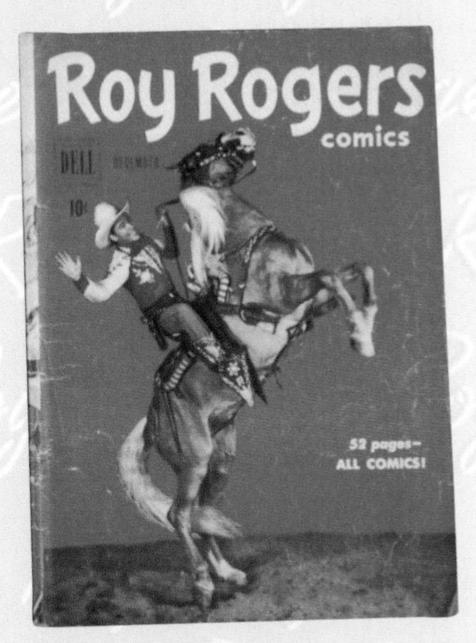

#36 December 1950. (R.L.)
Value $40-55

#38 February 1951. (R.L.)
Value $40-55

#40 April 1951. (R.L.)
Value $40-55

#45 September 1951. (R.L.)
Value $40-55

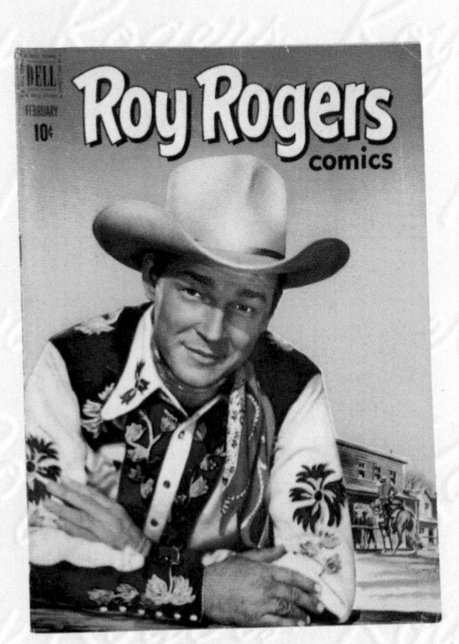

#50 February 1952. (R.L.)
Value $25-40

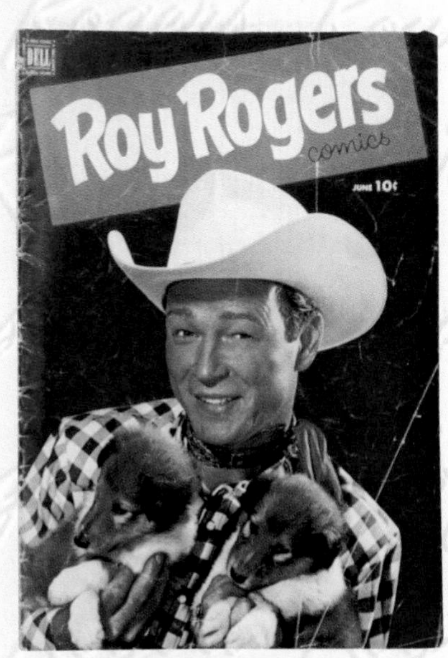

#54 June 1952. (R.L.)
Value $25-40

Roy Rogers Comic Album #2. Published by World Distributors, Ltd. 1953. Reprints of Roy Rogers Comics stories, "Mantrap on Longshot Mesa," "Man Hunt," "Clue of the Spur," "The Rustler on Goblin Hill," "The Rifleman of Boulder Wash," and "Canyon Burial." (G.S.)

Value $45-75

#62 February 1953. (R.L.)

Value $25-45

#63 March 1953. (R.L.)

Value $25-45

#66 June 1953. (R.L.)

Value $25-45

#68 August 1953. (R.L.)

Value $25-45

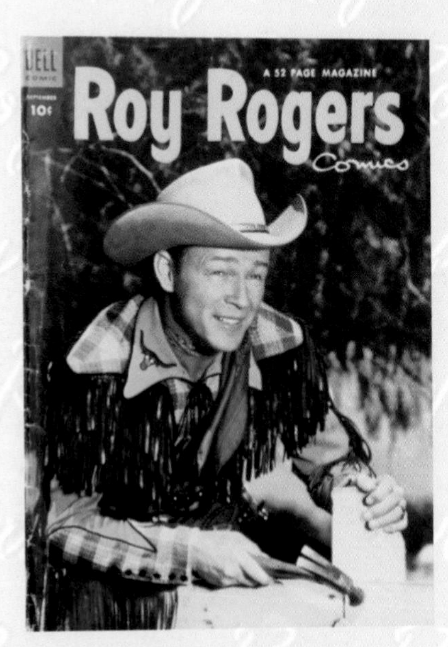

#69 September 1953. (R.L.)

Value $25-45

#72 December 1953. (R.L.)
Value $25-45

#73 January 1954. (R.L.)
Value $25-45

#75 March 1954. (R.L.)
Value $25-45

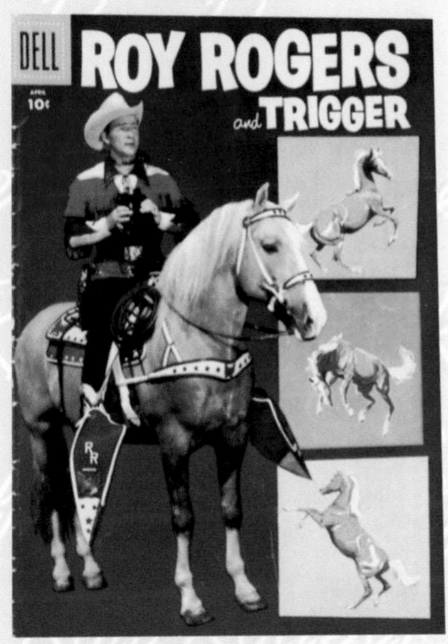

#100 April 1956. (R.L.)
Value $40-55

#103 July 1956. (R.L.)
Value $20-40

#106 October 1956. (R.L.)
Value $20-40

#111 March 1957. (R.L.)
Value $20-40

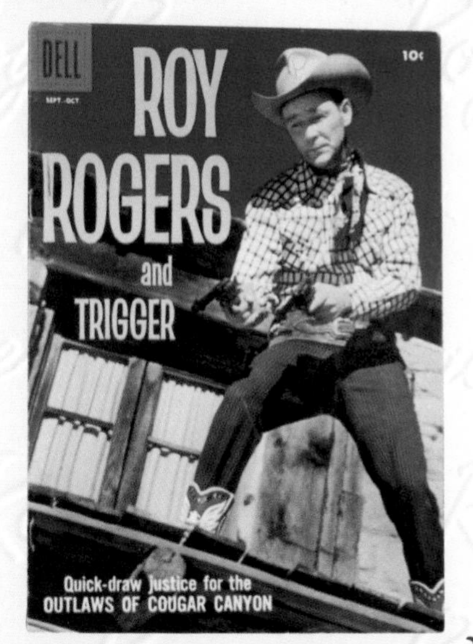

#127 September-October 1958. (R.L.)
Value $20-40

Original illustrations of Roy Rogers newspaper comic strip. Rare. 1950s. (G.S.)

Value $150-250 each

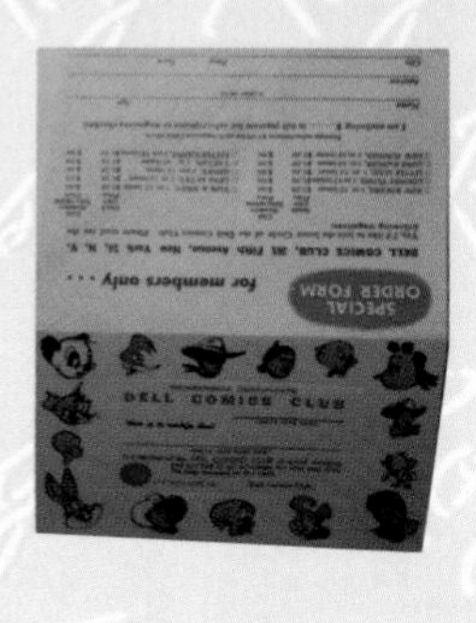

"Dell comics are Good comics" illustrated metal top sign from Dell Comics book stand with 8" x 10" color illustration of "Dell Family" and subscription order for comics. (G.S.)

Value $125-250

A complete collection of March of Comics featuring Roy Rogers. 1948-1963. 25 issues were published that included #5, 17, 35, 47, 62, 68, 73, 77, 86, 91, 100, 105, 116, 121, 131, 136, 146, 151, 161, 167, 176, 191, 206, 221, 236, and 250. These were free comics that stores such as Sears, Roebuck & Company and Poll-Parrot's children's shoes company gave to their customers. The comic first issued, #17 has a value of $125–200. Comic #250 is valued at $20-35 in C-8 to mint condition. (G.S.)

Dale Evans Comics issued by D.C. National Comics. Early 1950s.

Comic #12. (G.S.)

Value $75-125

Comic #10 (G.S.)

Value $175-225

#38 April 1944. (G.S.)
Value $950-2,400
#63 January 1945. (G.S.)
Value $450-600

Strip from 1950s Minneapolis Sunday Tribune newspaper. (G.S.)
Value $12-25

Dell Comics free offer for 8" x 10" color illustration of "Dell's Family," inside back cover of Roy Rogers comic book.

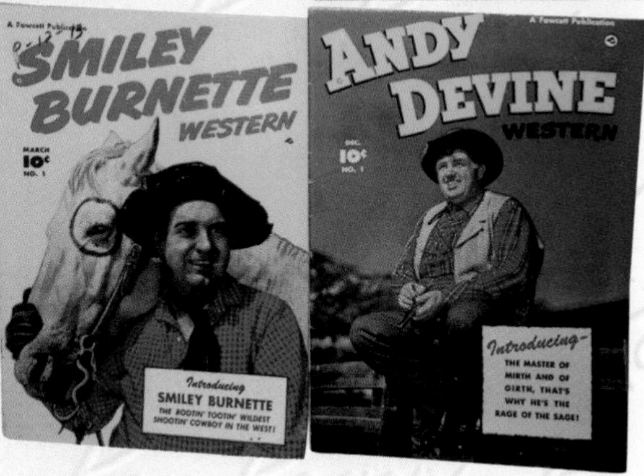

Western movie side-kicks of Roy Rogers featured in comic books:

Gabby Hayes Western #27 "Sourdough Gold." 1950s. Fawcett Publications. (G.S.)
Value $25-45

Gabby Hayes Western #36 "Rustlers' Rodeo." 1950s. Fawcett Publications. (G.S.)
Value $25-45

Smiley Burnette Western #1. 1950. (G.S.)
Value $250-300

Andy Devine Western #1. 1950. (G.S.)
Value $250-400

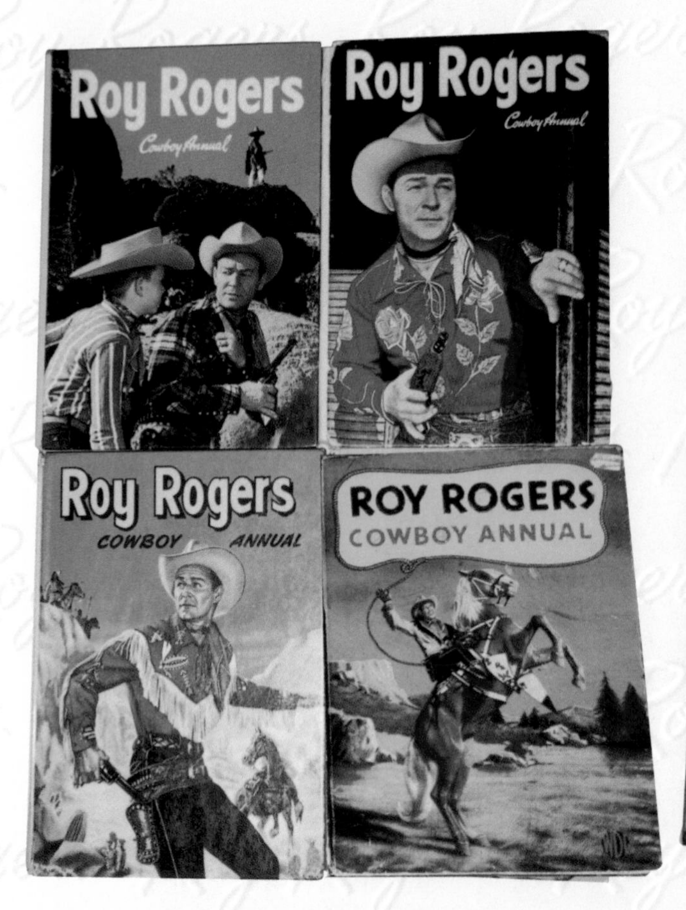

Roy Rogers Cowboy Annuals published by World Distributors, Ltd. Adprint/LTA Robinson, London. Eleven annuals were issued from 1951 to 1961 containing reprints of Roy Rogers Comics stories published by Dell Comics. Walt Howarth paintings of Roy and Trigger were used as front covers for some of the annuals like the cover of issue #3 depicting Roy on rearing Trigger. Values for all *Roy Rogers Cowboy Annuals* from C-8 to mint condition. (G.S.)

Value $45-75 each

Five Gabby Hayes small premium comics with mailer. Rare. (M.M.)

Value $85-135

Comic book printed in Mexico. 1952. Produced by E.N. Publishing Co. (C.Q.)
Value $10-20

Dale Evans, Queen of the West Comic. 1950s. Dell Publishing Co. (R.L.)
Value $25-50

Dale Evans, Queen of the West Comic. 1950s. Dell Publishing Co. (R.L.)
Value $25-40

March of Comics. 1950s. Photograph of Roy on front cover. Poll-Parrots giveaway. (G.S.)
Value $35-60

March of Comics. Early 1950s. Poll-Parrot's giveaway. Illustrated picture of Roy Rogers on cover. (G.S.)
Value $50-75 each

CHAPTER 10

CLOCKS AND WATCHES

Two major timepiece-manufacturing companies in the 1950s produced Roy Rogers and Trigger clocks, and Roy Rogers and Dale Evans watches. Ingraham Company, of Bristol, Conn., and Bradley Time Corporation manufactured very similar 40-hour animated wind-up alarm clocks. Clocks from both companies were the same 1 1/2" x 4" x 4 1/2" size and they both were made with enameled metal cases with brass or chrome frames around the clock face. These clocks came in colors of cactus (pale green), saddle (dark brown), sky (light blue) or desert sands (tan and ivory.) The faces of these clocks show an illustrated Roy and Trigger actually galloping across the beautiful desert landscape with the ticking of every second.

Clocks with Roy's name on them are valued at $150 more than clocks without his name. There are subtle differences in the illustrated scene on the clocks. Some Ingraham animated clocks feature Roy leaning forward in the saddle. On the face of the Bradley time clock, Roy sits straight in the saddle. The bases may also be different on these clocks.

Ingraham Company also produced a larger 5 1/8" x 5 1/8" animated alarm clock with a plastic case that came in the same colors as the smaller alarm clock. The Canadian version of the smaller alarm clock also was produced with a plastic case. This clock has "log" looking numbers and an illustrated mountain scene with trees rather than the American desert scene. These are very rare clocks.

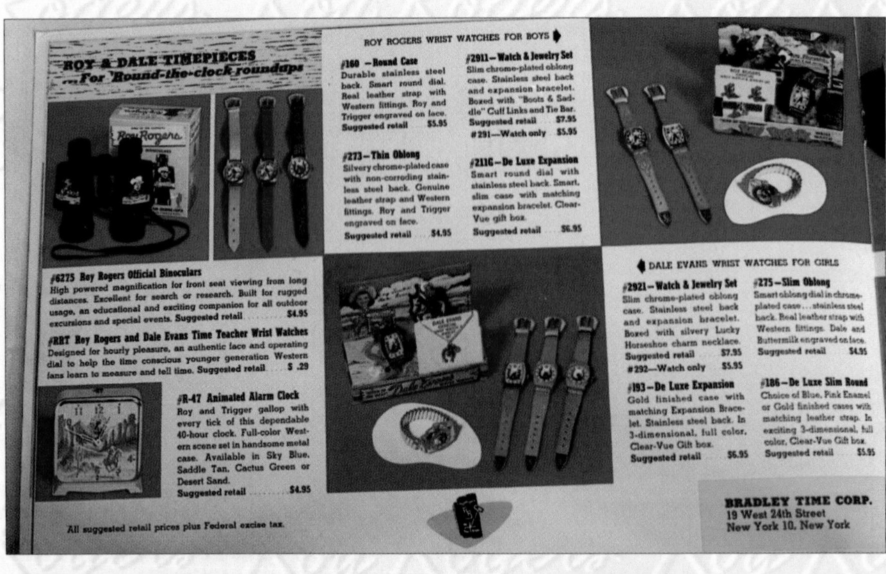

ROY & DALE TIMEPIECES
...For 'Round-the-clock roundups

ROY ROGERS WRIST WATCHES FOR BOYS

#160 —Round Case
Durable stainless steel back. Smart round dial. Real leather strap with Western fittings. Roy and Trigger engraved on face.
Suggested retail ... $5.95

#2911—Watch & Jewelry Set
Slim chrome-plated oblong case. Stainless steel back and expansion bracelet. Boxed with "Boots & Saddle" Cuff Links and Tie Bar.
Suggested retail ... $7.95
#291—Watch only ... $5.95

#273—Thin Oblong
Silvery chrome-plated case with non-corroding stainless steel back. Genuine leather strap and Western fittings. Roy and Trigger engraved on face.
Suggested retail ... $4.95

#211G—De Luxe Expansion
Smart round dial with stainless steel back. Smart, slim case with matching expansion bracelet. Clear-Vue gift box.
Suggested retail ... $6.95

#6275 Roy Rogers Official Binoculars
High powered magnification for front seat viewing from long distances. Excellent for search or research. Built for rugged usage, an educational and exciting companion for all outdoor excursions and special events. Suggested retail ... $4.95

#RRT Roy Rogers and Dale Evans Time Teacher Wrist Watches
Designed for hourly pleasure, an authentic face and operating dial to help the time conscious younger generation Western fans learn to measure and tell time. Suggested retail ... $ .29

#R-47 Animated Alarm Clock
Roy and Trigger gallop with every tick of this dependable 40-hour clock. Full-color Western scene set in handsome metal case. Available in Sky Blue, Saddle Tan, Cactus Green or Desert Sand.
Suggested retail ... $4.95

DALE EVANS WRIST WATCHES FOR GIRLS

#2921—Watch & Jewelry Set
Slim chrome-plated oblong case. Stainless steel back and expansion bracelet. Boxed with silvery Lucky Horseshoe charm necklace.
Suggested retail ... $7.95
#292—Watch only ... $5.95

#275—Slim Oblong
Smart oblong dial in chrome-plated case ... stainless steel back. Real leather strap with Western fittings. Dale and Buttermilk engraved on face.
Suggested retail ... $4.95

#193—De Luxe Expansion
Gold finished case with matching Expansion Bracelet. Stainless steel back. In 3-dimensional, full color, Clear-Vue Gift box.
Suggested retail ... $6.95

#186—De Luxe Slim Round
Choice of Blue, Pink Enamel or Gold finished cases with matching leather strap. In exciting 3-dimensional, full color, Clear-Vue Gift box.
Suggested retail ... $5.95

All suggested retail prices plus Federal excise tax.

BRADLEY TIME CORP.
19 West 24th Street
New York 10, New York

Bradley Time Co. retail merchants catalog advertisement for Roy Rogers and Dale Evans timepieces.
(Roy Rogers Museum)

Roy Rogers Time Teacher Watch, 1950s. (C.Q.)
Value $35-60

Roy Rogers and Trigger Ingraham Co. animated alarm clock with box and card. (R.L.)

Value $250-450

Roy Rogers and Trigger Ingraham Co. animated large light blue alarm clock with rare box. (G.S.)

Value $400-700

Roy Rogers and Trigger Ingraham Co. blue animated smaller clock with box. (G.S.)

Value $350-550

Roy Rogers and Trigger Bradley Time animated 40-hour alarm clock, sky color. (C.Q.)

Value $250-450

Roy Rogers alarm clock, recent manufacture. (C.Q.)

Value unknown.

Roy Rogers and Trigger Ingraham Co. animated 40-hour alarm clock. (C.Q.)

Value $250-450

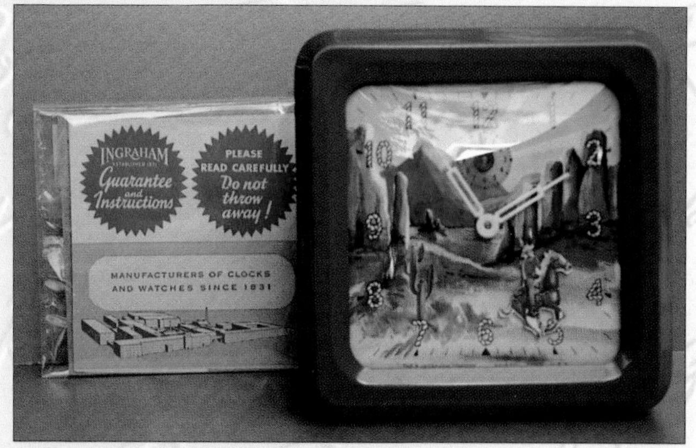

Large Roy Rogers and Trigger Ingraham Co. animated 40-hour alarm clock with warranty and instructions. Plastic dark brown case, 1952-1954. (C.Q.)

Value $350-550

Roy Rogers and Trigger Ingraham Co. animated 40-hour alarm clock, ivory color, with box, 1952-1954. (C.Q.)

Value $350-550

Roy Rogers and Trigger Bradley Time Canadian animated alarm clocks with red and white plastic cases, log-looking numbers, and illustrated mountain scenes. Very rare. (G.S.)

Value $400-600 with Roy Rogers name and box

Value $250-350 without the name and box

Roy Rogers wristwatch by Ingraham Co. with oblong face. Shows Roy and Trigger inside horseshoe design, chrome-plated case, buckle keeper, and leather band tip, 1952. (C.Q.)

Value $250-350, add $100 for original box

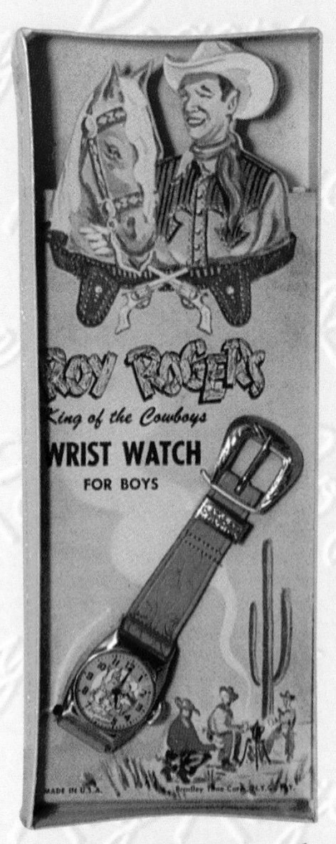

Roy Rogers wristwatch by Ingraham Co., thin round face, chrome-plated case, leather band, 1950s. (C.Q.)

Value $250-350, add $100 for original box

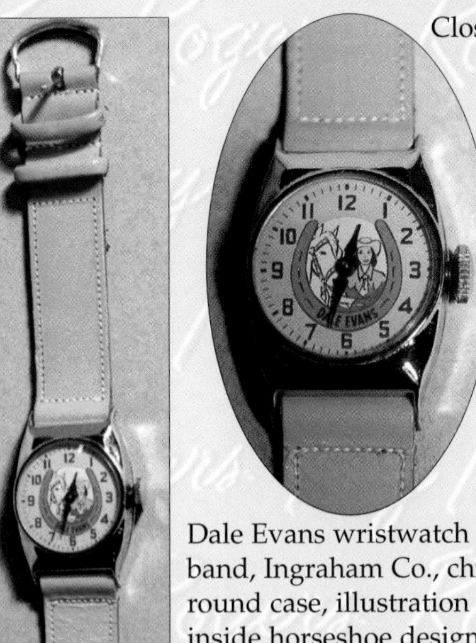

Close-up of Dale Evans wristwatch. (C.Q.)

Dale Evans wristwatch with leather band, Ingraham Co., chrome-plated round case, illustration of Dale inside horseshoe design on watch face, leather band, 1951. (C.Q.)

Value $200-350

Four 2" Roy Rogers pocket watches:

Top Center–Roy Rogers Ingraham Co. pocket watch, 1960 with a stop watch feature, face shows Roy on rearing Trigger. (C.Q.)

Value $250-400

Middle Center–Two Roy Rogers pocket watches, Ingraham Co. (C.Q.)

Value unknown

Middle bottom–Roy Rogers and Trigger pocket watch by Bradley Time, with stop watch feature, illustrated Roy portrait and silvered metal case, 1959. (C.Q.)

Value $250-400

Roy Rogers "King of the Cowboys" watch, 1950s. (C.Q.)

Value $35-60

Dale Evans expansion band wristwatch by Ingraham Co. with chrome-plated case and band, 1952. (C.Q.)

Value $300-450 watch and box

Value $150-300 watch only

CHAPTER 11

CLOTHING

Sears, Roebuck & Company and Macy's Department Store were two of the largest retail outlets for Roy Rogers and Dale Evans merchandise in the 1950s. Special displays like "The Roy Rogers Corral" were set up in different departments of these stores and offered all types of Roy Rogers and Dale Evans items. They sold complete Roy Rogers and Dale Evans western outfits, pants, shirts, sweaters, jackets, belts, boots, and hats with a wide array of designs, colors, and sizes. Parents would outfit their kids to look just like the "King of the Cowboys" and the "Queen of the West". Purchasers of Roy and Dale items were guaranteed that the item they were buying was of the best quality and value.

Roy Rogers leather vest with fringe and pockets, Roy Rogers brand in circled ropes on front. (R.L.)
Value $60-110

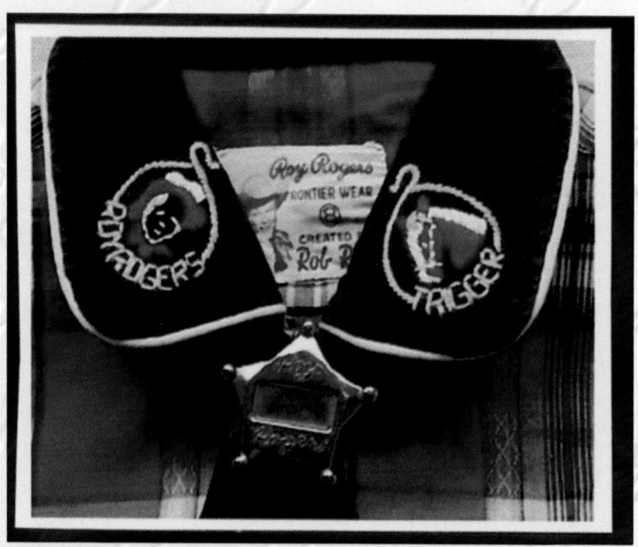

Roy Rogers Frontier Wear shirt by Rob Roy features two button snap cuffs and slash pocket. Roy Rogers and Trigger names on collar. Roy and Trigger necktie priced separately. (R.L.)
Value $50-80

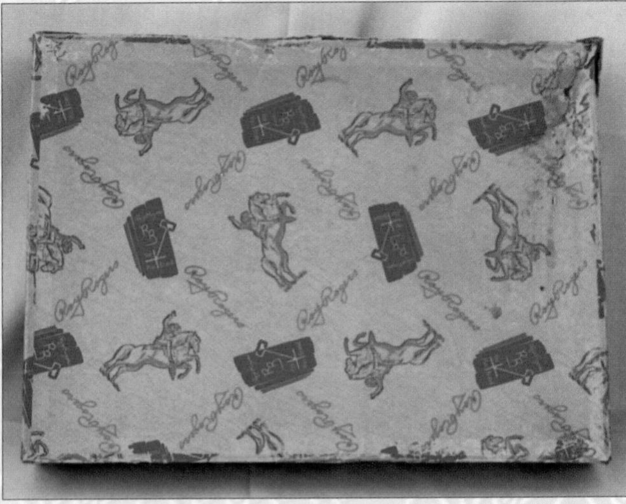

Roy Rogers boy's western suit. Shirt and pants by J Bar T, Inc. with original box from the early 1950s. (D.T.)

Value $150-250, as shown without box deduct $75

Roy Rogers and Dale Evans retail merchants catalog advertisements for clothes and accessories. (Roy Rogers Museum)

Little boy mannequin in Roy Rogers western attire complete with double holster and cap guns. (G.S.)

Roy Rogers cotton twill pants with mother of pearl pocket snaps, sold by Sears in the 1950s. (M.M.)

Value $75-150

Roy Rogers Riders coat and jeans with tags, 1950s. (G.S.)

Value $150-300

Dale Evans skirt and vest set. Notice Dale Evans' brand in butterfly logo, complete with very rare illustrated box, early 1950s. (B.W.)

Value $250-450

Roy Rogers "Frontier 45" western jeans with all original tags, 1950s. (Roy Rogers Museum)

Value $75-125

Roy Rogers denim jeans "11 Reasons why Roy Rogers denim jeans are the best" tag was stapled to jeans that were sold by Sears Roebuck stores. (C.Q.)

Value $15-30

Roy Rogers leather boy's chaps with metal conches, early 1950s. (Roy Rogers Museum)

Value $150-250

Roy Rogers boy's sweater. (R.L.)
Value $75-150

Roy Rogers boy's western shirt with studs, jewels, and fringe on yoke. (G.S.)
Value $125-200

Roy Rogers and Dale Evans 1950s western clothes by J Bar T, Inc. advertisements. (Roy Rogers Museum)

Dale Evans cowgirl outfit box by Yank Boy Play Clothes. (C.Q.)
Value $125-200

Roy Rogers Irwin Foster jacket advertisement, 1950s. (Roy Rogers Museum)

Original photo of prototype Roy Rogers leather jacket. (Roy Rogers Museum)

Roy Rogers and Pauker sweaters 8" x 10" original advertisement slick. (Roy Rogers Museum)

Roy Rogers denim jeans with "11 Reasons Why" tag, 1950s. (G.S.)

Value $75-125

Roy Rogers red leather jacket with rare original box. (G.S.)

Value $250-400

Roy Rogers complete seven-piece western outfit with original box. Set included chap-style pants, shirt, vest, belt, holster, tie, and metal clicker gun. Notice Roy Rogers "Pledge to Parents" tag. (B.W.)

Value $240-400

Roy Rogers and Trigger Official Cowboy Outfit made by Yank Boy Play Clothes and sold by Sears. This was originally an eleven-piece outfit with "Mountie" style hat, bat wing cotton twill chap-type pants, cotton shirt, tie, slide, lariat, belt, holsters, and clicker guns. (R.L.)

Value $250-400

Original photo of Roy Rogers prototype rainwear for children, 1950s. (Roy Rogers Museum)

Roy Rogers and Trigger corduroy jacket made by Irwin Foster Co., 1950s. (G.S.)

Value $150-300 with box

Dale Evans Official Cowgirl Outfit with box by Yank Boy. Set consists of skirt, vest, hat, and scarf. Rare. (M.M.)

Value $250-400

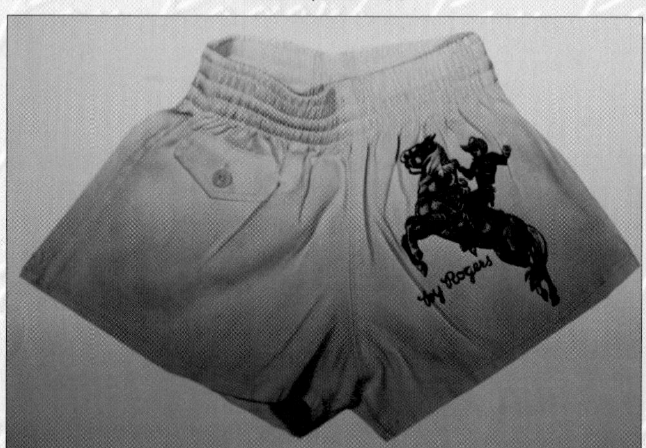

Original photos of Roy Rogers name and image with Trigger on prototype girl's shirt and shorts set. (Roy Rogers Museum)

Collection of Roy Rogers cowboy boots with original boxes; bunkhouse boots (slippers), and socks for children, 1950s. (G.S.)

Roy Rogers rodeo boots box. (C.Q.)

Value $125-200

Original ad photo for Roy Rogers and Dale Evans leather boots by Lone Star Boot Co., Inc. (Roy Rogers Museum)

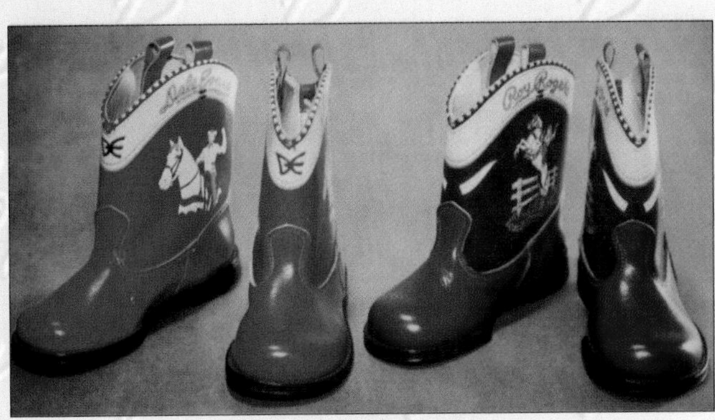

Roy Rogers and Dale Evans infant cowboy boot ad by Ettelbrick Shoe Company.

Roy Rogers children's boots with thunderbird designs. (C.Q.)

Value $150-250

Roy Rogers Cowboy Boots with name and image on boot upper sides made by Biltwel and sold by Sears. (G.S.)

Value $250-400

Rare and beautiful pair of Roy Rogers children's boots with original box and packing slip. Made by Tex-Tan of Texas in mid 1950s. Note perforated leather trim on heel and toe areas, colored leather inlay on boot uppers, and leather lacing on upper edges of boots. Roy Rogers name in gold stitching on boot pull tabs. These boots were made to look just like Roy's. (C.Q.)

Value $350-600

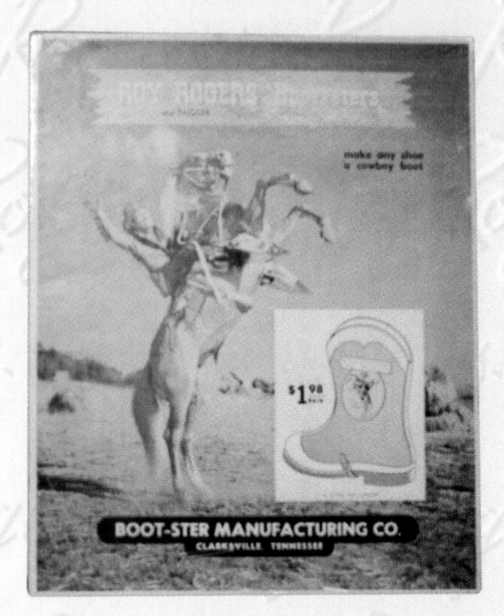

Roy Rogers boot-ster box. Boot-Ster Manufacturing Co. made children's boot-sters—cuffs worn around the ankles with shoes that gave the appearance of wearing cowboy boots in 1948. (C.Q.)

Value $75-125 box only

Roy Rogers Box for Bunkhouse Boots. (Roy Rogers Museum)

Value $75-125

Original photo of Roy Rogers and Dale Evans leather moccasins for adults. (Roy Rogers Museum)

Roy Rogers Tex-Tan cowboy boots box. (D.T.)

Value $90-150

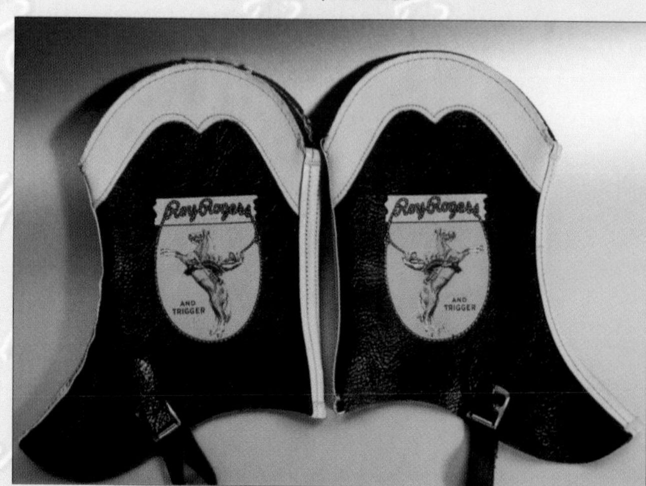

Roy Rogers and Trigger Boot-sters thunderbird design. (M.M.)

Value $125-200

Roy Rogers and Trigger Boot-sters made by Boot-Ster Manufacturing Co. in 1948. Rare. (C.Q.)

Value $125-200 each pair

Roy Rogers socks. (Roy Rogers Museum)
Value $50-75 each

Original photo of Roy Rogers prototype slippers. (Roy Rogers Museum)

Roy Rogers children's cowboy boot slippers. (C.Q.)
Value $125-200 each pair

Roy Rogers Classy Products spurs box. (C.Q.)
Value $150-200

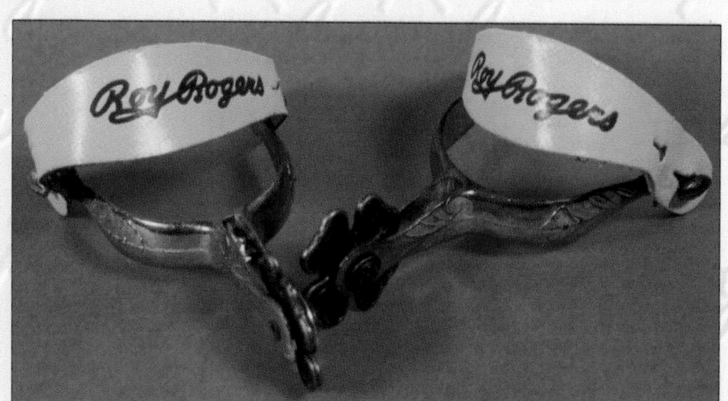

Roy Rogers Classy Spurs with Roy's name on leather straps and copper-plated, cloverleaf rowels. (C.Q.)
Value $175-250

Roy Rogers Leslie Henry Spurs with Roy's name and studs on leather straps, chrome-plated conches with gold-plated steer heads on left and right sides, nine-point star rowels, early 1950s. (C.Q.)
Value $175-250

Roy Rogers George Schmidt Spurs with copper-plated rowels and Roy Rogers copper letter on chrome-plated conches attached to black leather straps. (G.S.)
Value $250-450, with original box

Roy Rogers George Schmidt western spurs with box, Roy Rogers brand on metal conches, eight-point star rowels with box. (C.Q.)
Value $275-450

Roy Rogers Western spurs with Roy's name on leather straps and eight-point star rowels, 1950s. (C.Q.)
Value $150-225

Roy Rogers Classy embossed brown and white leather straps with Roy Rogers brand conches and eight-point star rowels. (C.Q.)
Value $175-250

Roy Rogers George Schmidt Spurs with square studs and metal Roy Rogers letters on straps and eight-point star rowels. (C.Q.)
Value $174-275

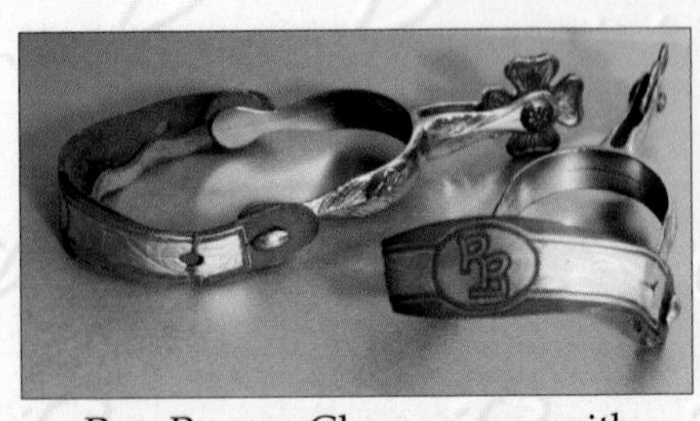
Roy Rogers Classy spurs with imprinted silver-colored Roy Rogers letters on red leather straps and copper-plated cloverleaf rowels. (C.Q.)
Value $175-250

Roy Rogers Western spurs with Roy's name in red letters on white strap and eight-point star rowels. (C.Q.)
Value $125-200

Dale Evans spurs. (C.Q.)
Value $275-450

Roy Rogers Sharp Shooter hat with box and instructions (listed under cap guns). (C.Q.)

Value $300-450

Three Roy Rogers and Dale Evans Sackman Brothers children's felt cowboy and cowgirl hats. (C.Q.)

Value $60-110 each

Roy Rogers and Trigger children's cowboy hats with adjustable rayon cord and chinstraps and laced felt brim. Note black straw hat in bottom left photo. (C.Q.)

Value $65-120

Roy Rogers child's cowboy hat. Black with brim whip-laced in yellow rayon and with yellow rayon chin cord; Roy on rearing Trigger imprinted on front of hat. (L.C.)

Value $75-125

Roy Rogers felt hat with unusual beaded hatband. Also has adjustable chinstrap and Roy Rogers brand. (B.W.)

Value $60-110

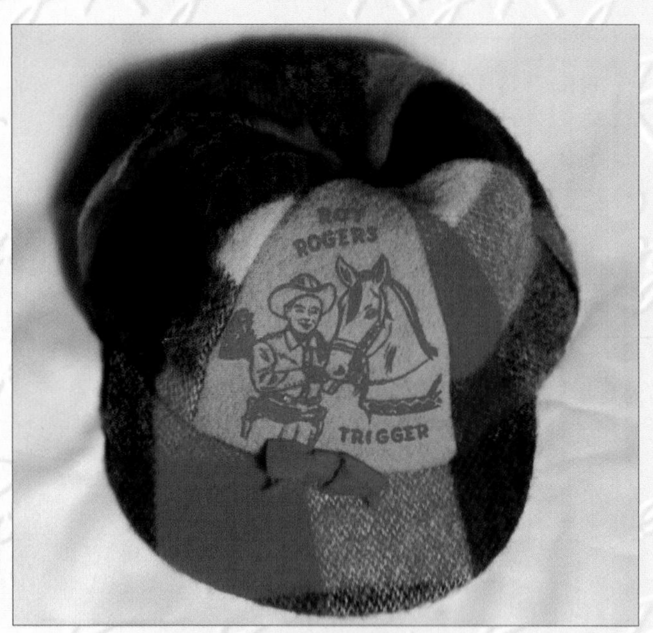

Roy Rogers Canadian wool cap with ear flaps, rare. (Roy Rogers Museum)

Value $200-300

Roy Rogers felt hat with whip-laced brim edge and Roy Rogers and Trigger's names on brown and yellow hatband. (Roy Rogers Museum)

Value $60-110

Very rare Gabby Hayes black hat. (M.M.)

Value $90-185

Roy Rogers beanie cap, rare. (B.W.)

Value $50-110

A collection of Roy Rogers and Trigger gauntlet-style leather gloves.
(G.S.)
Value $125-200 each

Roy Rogers and Trigger gauntlet suede gloves sold by Sears with leatherette fringed cuffs.
(C.Q.)
Value $125-220

Roy Rogers and Trigger gauntlet suede leather gloves with leatherette fringed cuffs. These gloves came in black, brown, green, and burgundy.
(C.Q.)
Value $125-225

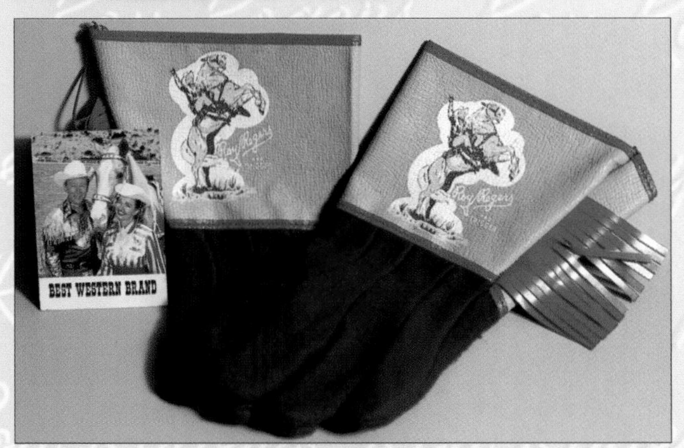

Roy Rogers and Trigger gauntlet suede leather gloves. Illinois Glove Company 1955-1956, tan leatherette cuff features picture of Roy on rearing Trigger, red plastic piping and fringe, Best Western brand name. (C.Q.)
Value $125-225

Roy Rogers gauntlet gloves 1955-1956 Illinois Glove Company. Lined deerskin with gold-colored metal signature and brand; came in russet and beige. (C.Q.)
Value $150-275

Roy Rogers suede leather gauntlet gloves with Roy's name and leather-studded overlay star and fringes on cuffs. (C.Q.)

Value $110-225

Roy Rogers leather brown gauntlet gloves by Illinois Glove Company. (C.Q.)

Value $150-275

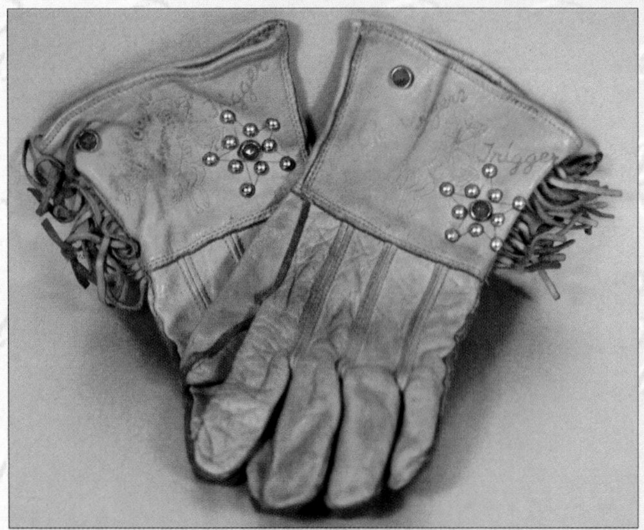

Roy Rogers deerskin leather gauntlet child's gloves, Illinois Glove Company. Cuffs are fringed with studs and red jewels; Roy Rogers and Trigger signature imprinted on cuffs. (C.Q.)

Value $150-275

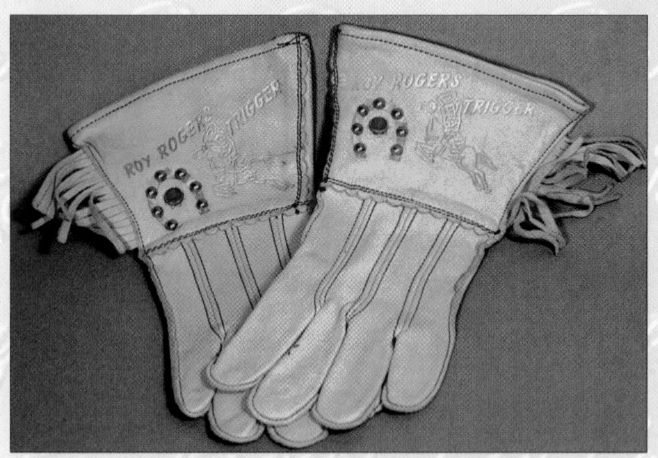

Roy Rogers and Trigger fringed deerskin gauntlet gloves. Roy Rogers and Trigger's name in gold-colored letters on cuffs. Studded horseshoe design on cuffs with red plastic jewel. Illinois Glove Company. (C.Q.)

Value $150-275

Roy Rogers leather cuffs with chrome-plated metal conches. (C.Q.)

Value $125-200

Original photograph taken in early 1950s of Roy Rogers holding belts that bear his name. (Roy Rogers Museum)

Roy Rogers sheriff's badge suspenders with five-point gold-plated star badge adjusters and gold-plated metal snaps. Hickok Manufacturing Company. Not on original card, 1949. (D.T.)

Value $85-125

Roy Rogers beaded leather belt with whip-laced edges, Roy's name done in black beads. (C.Q.)

Value $65-90

Roy Rogers sheriff's badge leather belt with fluted metal studs, Roy's signature, and gold-plated sheriff's badge-keeper. Hickok Manufacturing Co. Note: Belt matches sheriff's badge suspenders, 1949. (C.Q.)

Value $90-150

Roy Rogers leather belt with unusual studded leather and gold-plated buckle, raised relief of Roy's head with signature on hat's brim. (C.Q.)

Value $85-135

Roy Rogers black and white leather belt with raised relief metal bullet-shaped studs, red jewels, and Roy's signature in its original cardboard display holder. Rare. (G.S.)

Value $125-175

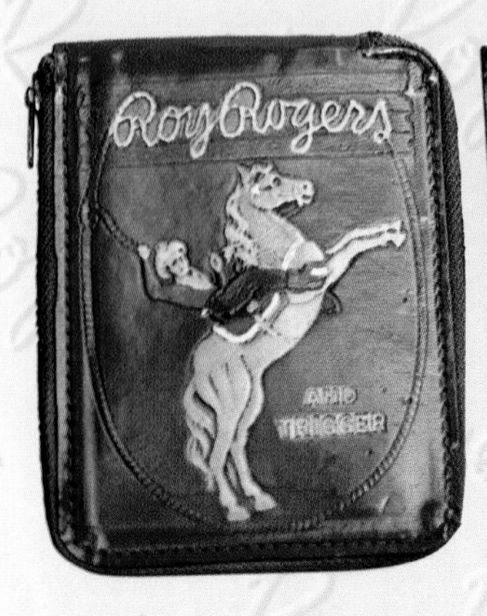

1. Roy Rogers leather wallet with two zippers on three sides. Roy waving on rearing Trigger in tinted color by Aristocrat Company in the late 1950s. (C.Q.)

Value $125-200

2.Roy Rogers billfold, 3 1/2″ x 4 1/2″ closed. Aristocrat Company. (C.Q.)

Value $125-200

Roy Rogers leather wallet with original box made in 1950 by Hickok Manufacturing Company, embossed with Roy and Trigger hand-tinted color. Came with Roy Rogers Riders Club membership card inside. (C.Q.)

Value $150-250

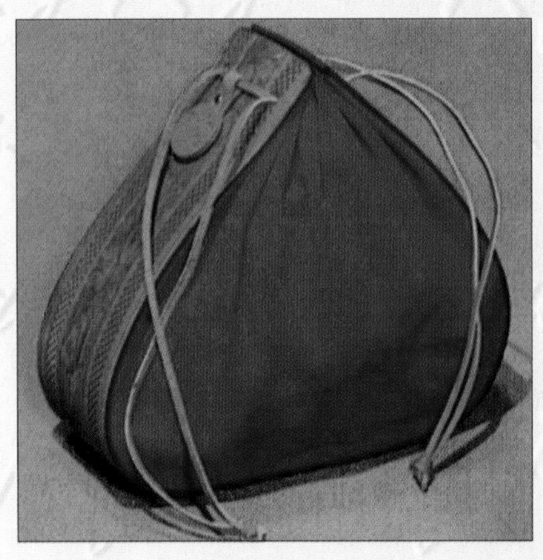

Dale Evans saddle leather and suede stirrup bags, 1950s. Available in red, brown, blue, tan, white, and black colors, manufactured in three sizes by Angelus Souvenir Manufacturing. (Roy Rogers Museum)

Value $75-150

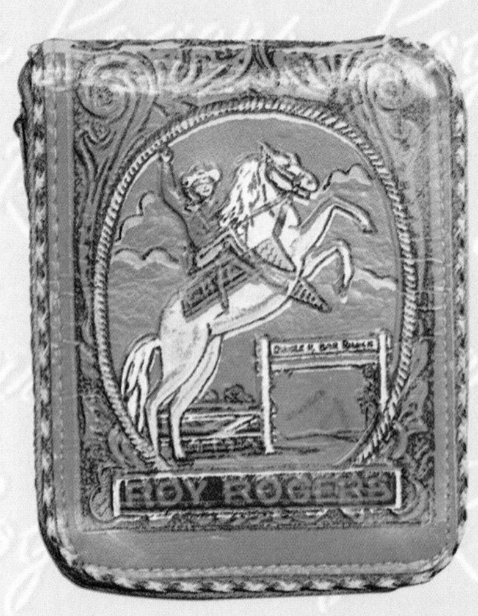

Roy Rogers sunset tan calfskin wallet. Roy on rearing Trigger encircled by lariat. Embossed and hand-tinted, two-toned braid edging with zipper on three sides 3 1/2″ x 4 1/2″ and sold by Sears, 1952. (C.Q.)

Value $125-200

Collection of Roy Rogers wallets and billfolds manufactured by Aristocrat Leather Products, Inc., 1950s. (Roy Rogers Museum)

Value $125-200

Roy Rogers and Trigger white satin tie with red illustrated names and image, Western Art Manufacturing Company. (C.Q.)
Value $45-75

Roy Rogers and Trigger satin tie with metal cuff slide and original box; came in blue, yellow, red, green, white, and black colors, Western Art Manufacturing Company, 1953. (M.M.)
Value $100-150

Roy Rogers and Trigger kerchief and scarf slide in original box, Western Art Manufacturing Company, 1950s. (R.L.)
Value $100-200

Roy Rogers and Trigger satin tie with metal sheriff's badge slide and original acetate plastic window box, Western Art Manufacturing Company, 1954. (Roy Rogers Museum)
Value $100-150

Roy Rogers Western "Blinking Bull" bolo tie. Bull's eyes light up when cord is pulled, on original display card. (C.Q.)
Value $75-125

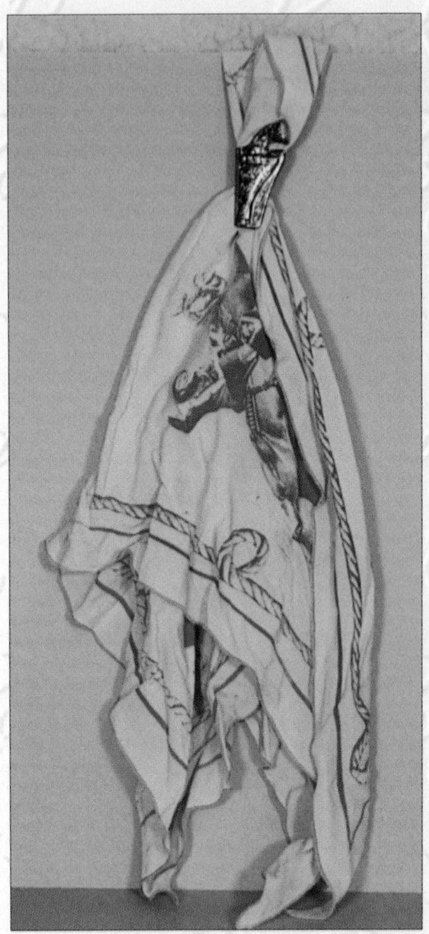

Roy Rogers and Trigger cream-colored 24″ square neckerchief with Roy Rogers metal gun and holster slide. (C.Q.)

Value $125-185

Roy Rogers and Trigger bandana scarf with metal cuff slide. (C.Q.)

Value $75-125

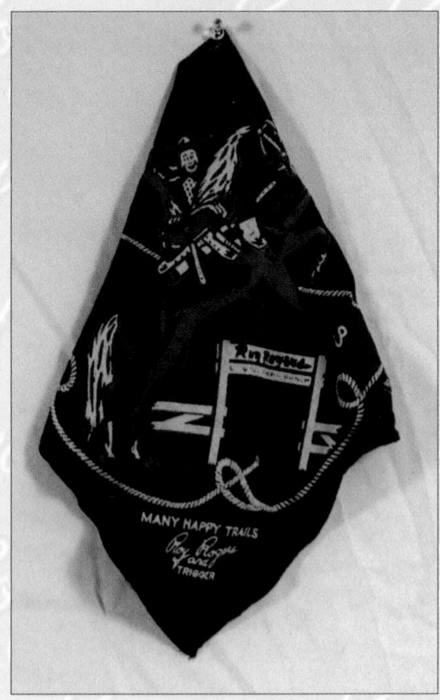

Roy Rogers "Many Happy Trails" 14″ square cotton neckerchief, white and green illustrations on blue background. (C.Q.)

Value $35-65

Dale Evans copper-plated cowboy hat scarf holder, Thrift Novelty Co., Inc., early 1950s. Original price was 29¢. (C.Q.)

Value $30-65

Roy Rogers raised relief metal gun and holster neckerchief slide. (C.Q.)

Value $50-75

Roy Rogers King of the Cowboys 24″ square red and golden yellow silk neckerchief with metal Roy Rogers cuff slide by Western Art Manufacturing Company with image of Roy on rearing Trigger, 1950s. (C.Q.)

Value $125-185

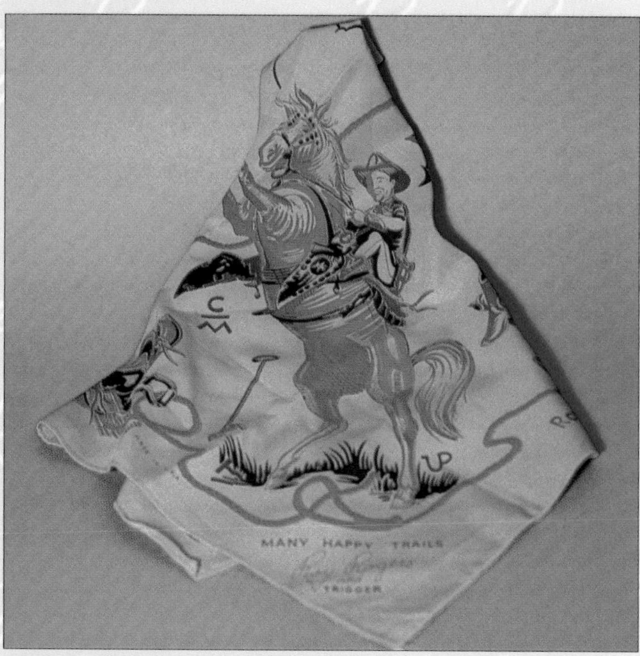

Roy Rogers and Trigger silk cream kerchief. (R.L.)
Value $60-120

Roy Rogers silk 24" square red and yellow neckerchief. Illustrated with Roy on rearing Trigger and Roy's facsimile signature. (C.Q.)
Value $60-120

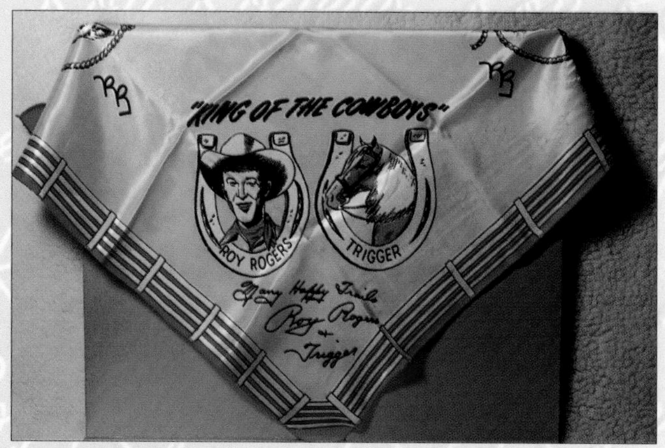

Roy Rogers "King of the Cowboys" silk neckerchief, Western Art Manufacturing Company. White fence on borders, illustrated images of Roy and Trigger, 24" square, yellow on red design, also came in red on yellow, 1940s. (C.Q.)
Value $60-120

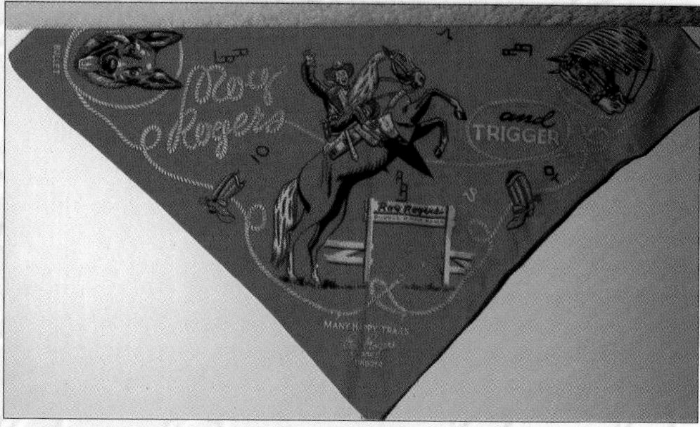

Roy Rogers "Many Happy Trails" 14" square cotton neckerchief, white and black illustrations on red background, 1950s. (C.Q.)
Value $35-65

Roy Rogers King of the Cowboys 14" square red cotton neckerchief, black printed images of Roy and Trigger framed in horseshoes. (C.Q.)
Value $35-75

Roy Rogers and Trigger silk cream-colored 24" square neckerchief. Rope design on borders. Illustrated Roy on rearing Trigger and various brands. (C.Q.)
Value $60-120

CHAPTER 12

Fan Club Memorabilia

Filmed in 1938, Roy Rogers' first movie, *Under Western Stars*, was voted Best Western of the Year and gave Roy star status at Republic Pictures. Positive reviews in movie magazines and a vigorous schedule of personal appearances made him a huge hit with the public almost overnight.

In his first year of stardom, Roy received 28,000 pieces of fan mail weekly and had to hire four women to answer the letters and send out autographed photographs. By 1946, there were over 700 Roy Rogers Fan Clubs in the United States and hundreds more in Canada and Great Britain. Roy received over one million fan letters a year by 1947 as his popularity kept increasing.

Roy Rogers Fan Clubs continued to grow as Roy Rogers Enterprises licensed merchandise flooded retail stores with millions of items and blanketed magazines, newspapers, and catalogs with Roy Rogers brand merchandise advertisements. More parents and children became fans with the 1948 radio broadcast of *The Roy Rogers Show* sponsored by General Foods, the makers of Quaker Oats Cereal.

The Roy Rogers Riders Club was formed in 1948 and had over one million seven hundred thousand children join in its first three months of operation. Two years later, there were three thousand Roy Rogers Riders Clubs. A Roy Rogers Riders Club in Oregon made the world's largest birthday cake celebrating Roy's birthday in 1950. The cake weighed a ton and was as big as a ping pong table.

With the advent of more Roy Rogers western movies, public appearances, television shows, plus massive merchandising of Roy Rogers brand items, his popularity soared and his fan clubs flourished. The Roy Rogers Fan Club in London with fifty thousand members was the largest fan club in the world. Roy Rogers fans continue to honor their favorite western hero, and celebrate his life and legacy as members of the Roy Rogers Riders Club and the Roy Rogers–Dale Evans Collectors Association.

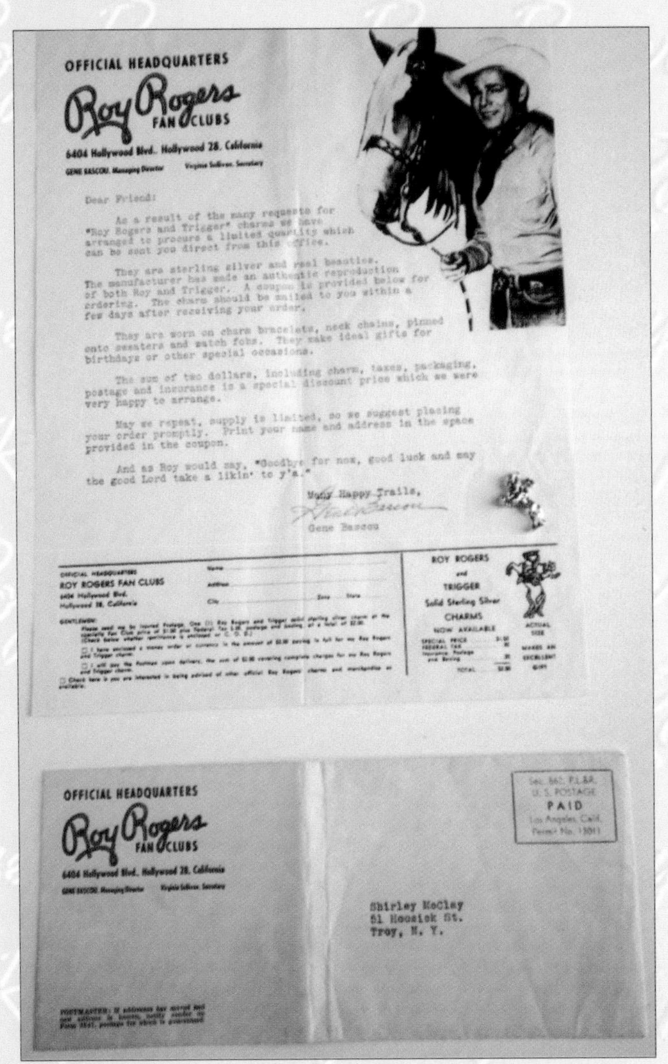

OFFICIAL HEADQUARTERS
Roy Rogers FAN CLUBS
6404 Hollywood Blvd., Hollywood 28, California
GENE BASCOU, Managing Director   Virginia Sullivan, Secretary

Dear Friend:

As a result of the many requests for "Roy Rogers and Trigger" charms we have arranged to procure a limited quantity which can be sent you direct from this office.

They are sterling silver and real beauties. The manufacturer has made an authentic reproduction of both Roy and Trigger. A coupon is provided below for ordering. The charm should be mailed to you within a few days after receiving your order.

They are worn on charm bracelets, neck chains, pinned onto sweaters and watch fobs. They make ideal gifts for birthdays or other special occasions.

The sum of two dollars, including charm, taxes, packaging, postage and insurance is a special discount price which we were very happy to arrange.

May we repeat, supply is limited, so we suggest placing your order promptly. Print your name and address in the space provided in the coupon.

And as Roy would say, "Goodbye for now, good luck and may the good Lord take a likin' to y'a."

Many Happy Trails,

Gene Bascou

OFFICIAL HEADQUARTERS
ROY ROGERS FAN CLUBS
6404 Hollywood Blvd.
Hollywood 28, California

Name
Address
City   Zone   State

GENTLEMEN:

ROY ROGERS and TRIGGER Solid Sterling Silver CHARMS NOW AVAILABLE
ACTUAL SIZE
SPECIAL PRICE
FEDERAL TAX
MAKES AN EXCELLENT GIFT

OFFICIAL HEADQUARTERS
Roy Rogers FAN CLUBS
6404 Hollywood Blvd., Hollywood 28, California

Sec. 562, P.L.&R.
U. S. POSTAGE
PAID
Los Angeles, Calif.
Permit No. 13011

Shirley McCley
51 Hoosick St.
Troy, N. Y.

Roy Rogers Fan Club letter to members with order form for a sterling silver charm, shown with charm and original envelope. (G.S.)

Value $75-125

THIS CERTIFIES THAT
RON LENIUS
IS A MEMBER IN GOOD STANDING OF THE
ROY ROGERS RIDERS CLUB
AND IS EXPECTED TO FOLLOW THE RULES FAITHFULLY
Roy Rogers
KING OF THE COWBOYS

ROY ROGERS RIDERS RULES

1. Be neat and clean.
2. Be courteous and polite.
3. Always obey your parents.
4. Protect the weak and help them.
5. Be brave but never take chances.
6. Study hard and learn all you can.
7. Be kind to animals and care for them.
8. Eat all your food and never waste any.
9. Love God and go to Sunday School regularly.
10. Always respect our flag and our country.

Many Happy Trails Roy Rogers & Trigger

Roy Rogers Riders Club membership certificate, recent. (R.L.)

Roy Rogers Riders Club card with "Lucky Piece" coin and note card from Roy inviting a friend to join the club, 1950s. (M.M.)

Value $35-60

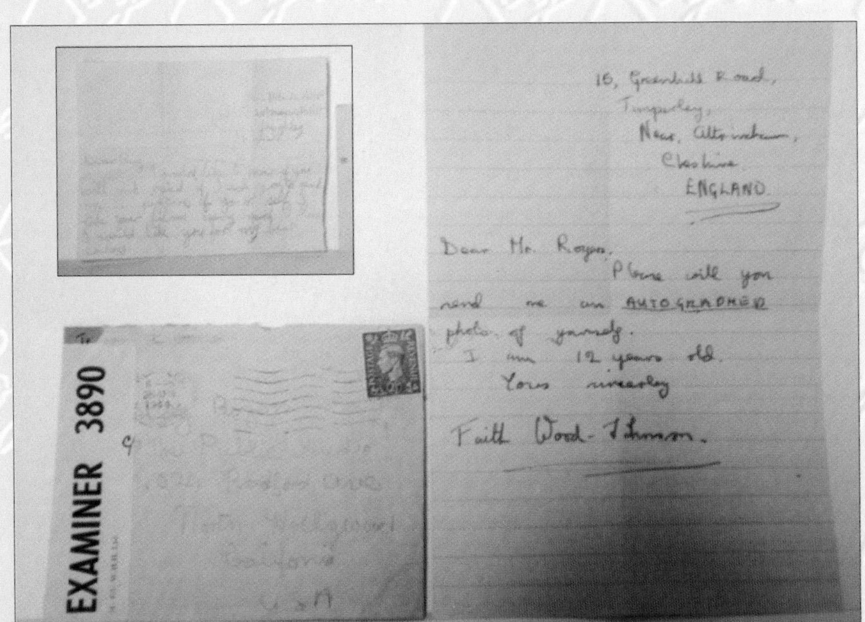

16, Greenhill Road,
Timperley,
Near Altrincham,
Cheshire
ENGLAND

Dear Mr. Rogers,
Please will you send me an AUTOGRAPHED photo of yourself.
I am 12 years old.
Yours sincerely
Faith Wood-Johnson.

EXAMINER 3890

Letters from Roy Rogers fans in England with original envelopes, rare, 1940s. (M.M.)

Value unknown.

Roy Rogers Riders Club membership kit, including comic book, button, card, mailer, cap gun with holster, order form, and photograph of Roy and Trigger, 1952. (M.M.)

Value $125-200

Roy Rogers Riders Club membership card, recent. (C.Q.)

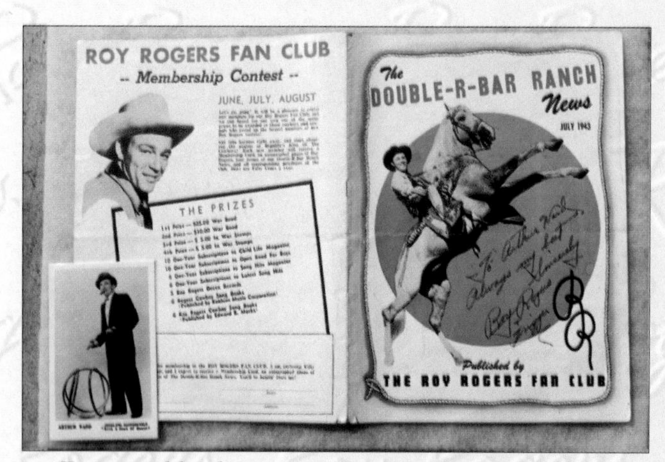

Front and back cover of the Double-R-Bar Ranch News club magazine published by The Roy Rogers Fan Club in 1943. Note: Roy Rogers and Trigger autograph. (G.S.)

Value $100-175

These are the dates and cities, where
ROY ROGERS,
"King of the Cowboys" will appear.

DATE	CITY
November 4	ST. JOSEPH, MISSOURI
November 5	DES MOINES, IOWA
November 6	MINNEAPOLIS, MINNESOTA
November 7	ROCHESTER, MINNESOTA
November 8	DULUTH, MINNESOTA
November 9	FARGO, NORTH DAKOTA
November 10	BISMARCK, NORTH DAKOTA
November 11	ABERDEEN, SOUTH DAKOTA
November 12	MITCHELL, SOUTH DAKOTA
November 13	SIOUX CITY, IOWA
November 15	CEDAR RAPIDS, IOWA
November 16	MILWAUKEE, WISCONSIN
November 17	GRAND RAPIDS, MICHIGAN
November 18	KALAMAZOO, MICHIGAN
November 19	SAGINAW, MICHIGAN
November 21	KOKOMO, INDIANA
November 22	SPRINGFIELD, ILLINOIS
November 23	BURLINGTON, IOWA
November 24	DAVENPORT, IOWA
November 25	SOUTH BEND, INDIANA
November 26	TROY, OHIO
November 27	MARION, OHIO
November 28	TOLEDO, OHIO
November 29	HUNTINGTON, WEST VIRGINIA
November 30	LOUISVILLE, KENTUCKY
December 1	NASHVILLE, TENNESSEE
December 2	MEMPHIS, TENNESSEE
December 3	BIRMINGHAM, ALABAMA

Letter sent to fan club members listing the dates and cities where Roy will appear.(R.L.)

Value $20-35

Roy Rogers Riders Club membership kit, includes original mailer, Post Cereal offer for cap gun with holster set, order form, card, button, and colored photograph of Roy and Trigger, 1950s. (G.S.)

Value $125-200

CHAPTER 13

INTERIOR FURNISHINGS

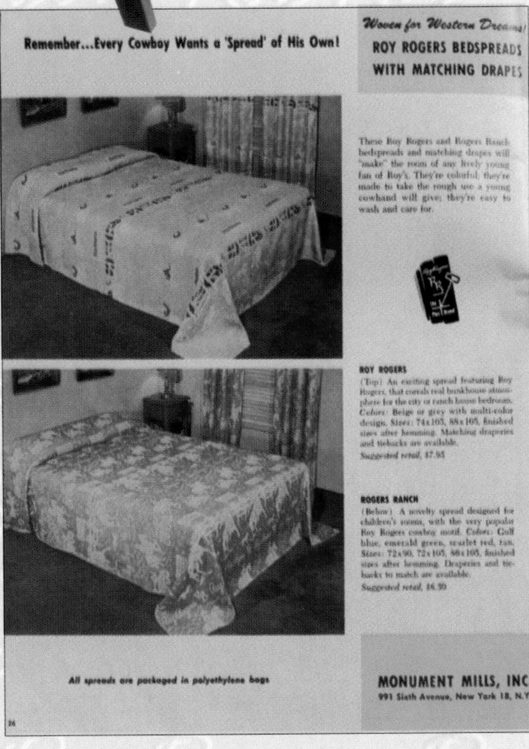

Retail merchants catalog advertisements for Roy Rogers blankets, bedspreads, curtains, towels, and wash mitts manufactured by Fieldcrest Mills and Monument Mills, Inc. (Roy Rogers Museum)

Roy Rogers Double-R-Bar Ranch 64" x 84" bedspread manufactured by Monument Mills, Inc. 1950s. (R.L.)
Value $125-200

Happy Trails Roy Rogers and Dale Evans bedspread with matching pillows. (C.Q.)

Value unknown

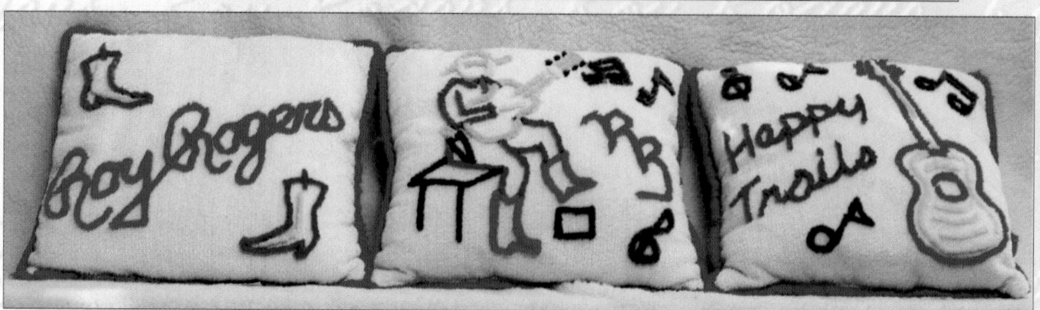

Three-Roy Rogers pillows. (C.Q.)

Value unknown

Roy Rogers pillows manufactured by Dakota in limited editions. Recent. (C.Q.)

Value $45-75 each

Roy Rogers and Trigger terry cloth towels sold by Sears, 1950s. (C.Q.)

Value $25-45 each

Original photo of Trigger's Hi Stepper. Child's stepping stool placed in front of bathroom sink for a child to use when washing hands, 1950s. (Roy Rogers Museum)

Value unknown

Original photo of Roy Rogers Televiewer, a wooden bench to sit on while watching television, 1950s. (Roy Rogers Museum)

Roy Rogers and Dale Evans child's card table with chairs. Leatherette-covered chair seats and tabletop, painted metal chairs and table. (B.W.)

Value $350-600

Roy Rogers child's wood rocking chair. (C.Q.)

Value $450-750

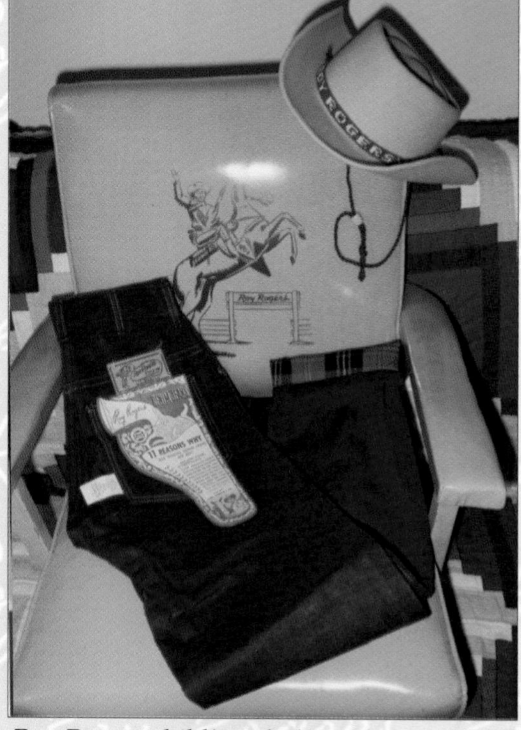

Roy Rogers child's upholstered chair. (G.S.)

Value $750-1,200

Original photo of prototype Roy Rogers upholstered child's chair, 1950s. (Roy Rogers Museum)

Roy Rogers and Trigger rare and unusual large lamp with wood base, inner shape revolves showing different scenes in cutout area of outer lampshade. (B.W.)

Value $500-750

Roy Rogers and Trigger painted 15″ plaster lamp with cream-colored illustrated shade. (B.W.)

Value $350-550

Roy Rogers and Trigger extremely rare large lamp. What a beauty! (B.W.)

Value $950-1,200

Roy Rogers and Trigger plaster lamp, 1950s, lamp base 8 1/2″ high with Roy on rearing Trigger and illustrated scene on lampshade. (C.Q.)

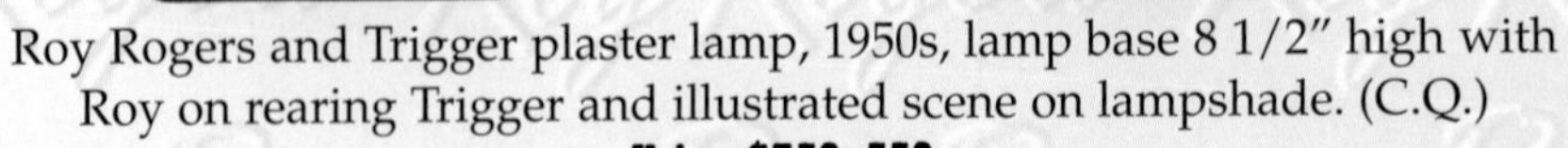

Value $350-550

"Happy Trails" Roy Rogers and Trigger ceramic plate 6" diameter with fluted design. (C.Q.)
Value $25-60

Roy Rogers and Trigger on 6" diameter plate with Roy twirling lariat by Universal Co., 1950s. (C.Q.)
Value $75-125

Roy Rogers on rearing Trigger 9" diameter plate by Universal Co. 1950s. (C.Q.)
Value $90-135

Roy Rogers and Dale Evans collector plates early 1990s by The Hamilton Collection. (C.Q.)
Value $35-60 each

Roy Rogers and Trigger ovenproof 6" diameter bowl by Universal Co. 1950s. (C.Q.)
Value $75-125

Roy Rogers and Trigger drinking glass. (C.Q.)
Value unknown.

Roy Rogers and Trigger drinking glass front and back views, 1952 by Federal Glass Company. Shows measurements in back that encourages children to drink all of their milk. (C.Q.)
Value $65-115

Roy Rogers Restaurant paper cup, 1980–1990s. (C.Q.)
Value $7-12 each

Roy Rogers and Trigger ceramic mugs, 1950s. (C.Q.)
Value $65-100 each

Roy Rogers ceramic cups 1980s. (C.Q.)
Value $8-15 each

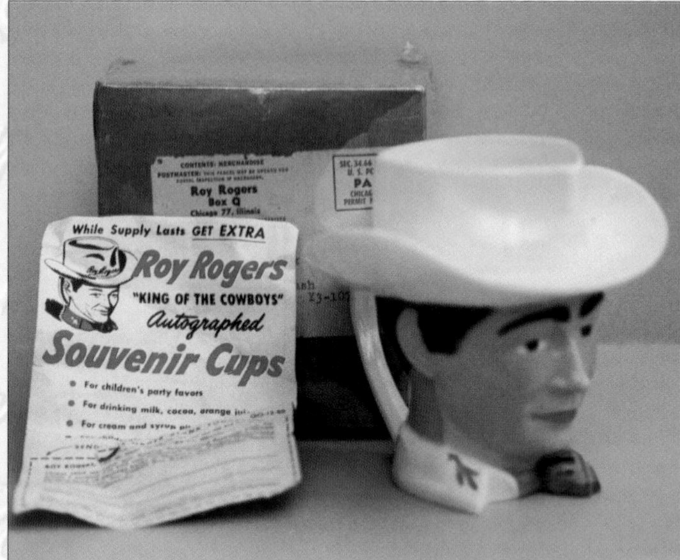

Roy Rogers Quaker Oats plastic souvenir cup. As shown with box and order form. (C.Q.)
Value $75-150

Roy Rogers paper napkins, 1950s. (C.Q.)
Value $10-18 each

Roy Rogers and Trigger ceramic cookie jar by McME Productions, limited edition 11 1/2" tall with Roy's signature on back of collar, 1994. (C.Q.)
Value $150-250

CHAPTER 14

MAGAZINES

Thousands of Roy Rogers and Dale Evans publicity photographs were taken from the late 1930s through the 1990s. These photos were used for merchandising, event promotion, books, comics, newspaper articles, and magazine covers. Generally, the older the magazine, the greater its value. Condition is a prime factor when considering the value of very fragile paper items, such as magazines. Pages yellow with age, creases or rips in the pages or cover, and the overall appearance of the cover affect the value of any paper collectible.

Roy Rogers featured on the cover of early 1950s *Western Stars* magazines.
Value $35-75 each

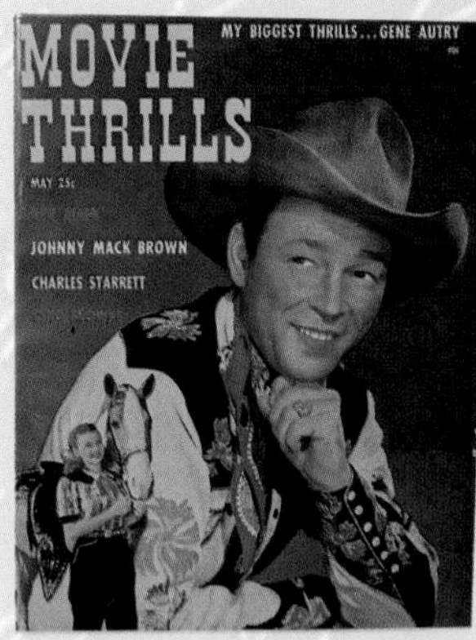

Roy Rogers on the cover of *Movie Thrills* magazine 1940s. (C.Q.)
Value $45-80

Top row:
1.*The Modern Horseman*, 1950s. (G.S.) **Value $20-35**
2.*Our Dogs*, 1940s, rare. (G.S.) **Value $75-115**
3.*Conservation Bulletin*. (G.S.) **Value $25-60**
Bottom row:
1.*Cue* magazine, 1950s. (G.S.) **Value $25-35**
2.*Cowboy Movie Thrillers*, 1950s. (G.S.) **Value $25-45**
3.*Possibilities* magazine, 1980s. (G.S.) **Value $15-30**

Gabby Hayes on the cover of *TV Today Magazine*. (M.M.)
Value $40-69

Roy Rogers featured on the covers of *Young America* magazines published in the early 1940s. (G.S.)
Value $45-90 each

Roy Rogers on the cover of *Country America* magazine, 1992. (R.L.)
Value $5-15

Roy Rogers and Dales Evans featured on a variety of magazine covers. (G.S.)

Value $40-60 each

Roy Rogers and Dale Evans featured on a variety of magazine covers. (G.S.)

Value $25-75 each

Roy Rogers on the cover of *Century Crusader* magazine, autographed, 1947. (C.Q.)

Value $50-75

Roy Rogers and Dale Evans on the cover of *Christian Leadership Bulletin*, 1957. (R.L.)

Value $8-15

Roy Rogers on the cover of *Scene* magazine, 1975. (C.Q.)

Value $25-50

Roy Rogers featured on the covers of 1940s and 1950s movie magazines. (G.S.)

Value $45-90 each

Roy Rogers and Dale Evans featured on late 1940s issues of *Movie Life* magazine. (G.S.)

Value $45-90 each

Roy Rogers and rearing Trigger on the cover of *Motor Transportation* magazine, 1958. (G.S.)

Value $45-90

Roy Rogers and Dale Evans featured on the cover of *The Saturday Evening Post*, 1980. (C.Q.)

Value $25-40

Roy Rogers on the cover of *California* magazine, 1982. (R.L.)

Value $20-35

Roy Rogers and Gabby Hayes featured on television guide magazines. (G.S.)

Value $40-60 each

Roy Rogers on the cover of *Fan Club Bulletin*, 1982. (C.Q.)

Value $12-18

Roy Rogers on the cover of *Gems and Minerals* magazine, 1968. (R.L.)

Value $35-65

Roy Rogers featured on the cover of 1940s magazines. (G.S.)

Value $45-90 each

Top row:

1. *Liberty* magazine, early 1950s (G.S.) **Value $45-90**

2. *Movie Fan*. (G.S.) **Value $45-90**

3. *Movie Fan*. (G.S.) **Value $35-75**

Bottom row:

1. *Liberty Magazine*, 1950s. (G.S.) **Value $35-75**

2. *Movie Play Western Edition*. (G.S.) **Value $35-75**

3. *Movies Magazine*, early 1950s. (G.S.) **Value $35-75**

Top row: Roy Rogers rodeo programs, 1950s. (G.S.)

Value $35-85 each

Bottom row: Roy Rogers and Dale Evans on the covers of *Rodeo* magazines, 1950s. (G.S.)

Value $45-100 each

Rare Roy Rogers rodeo program, 1945. (R.L.)

Value $75-125

Roy Rogers and Dale Evans featured on *Texas Monthly* magazine, 1978. (C.Q.)

Value $12-25

Roy Rogers and Trigger photo on Los Angeles Transit Lines bus ticket advertising, Sheriffs Rodeo, rare, 1948. (C.Q.)

Value Unknown.

Top row: Roy Rogers featured on rodeo programs early 1940s. Rare. (G.S.)

Value $75-125 each

Bottom row: Roy Rogers featured on the covers of *Rodeo* magazines, 1950s. (G.S.)

Value $50-90 each

Collection of Roy Rogers and Dale Evans on the covers of rodeo programs. (G.S.)

Value $25-100 each

Roy Rogers on the cover of Sheriff's Rodeo program, 1950. (R.L.)

Value $25-60

Roy Rogers on Madison Square Garden Rodeo poster, 13" x 21" and extremely rare. (G.S.)

Value $200-350

Roy Rogers on the cover of *Fan Fare* magazine, 1943 issue. (C.Q.)

Value $45-90

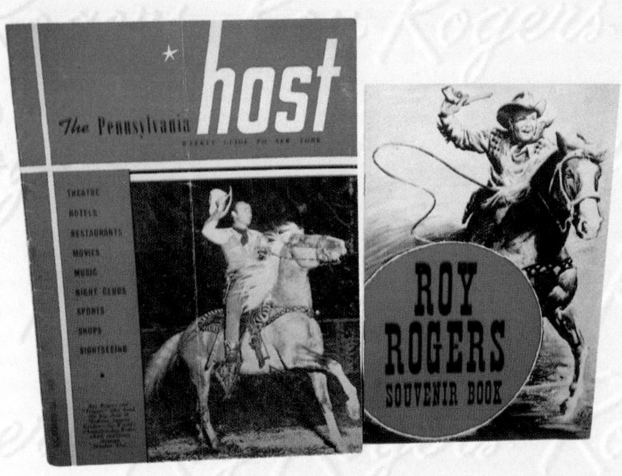

1. Roy Rogers and Trigger on the cover of *The Pennsylvania Host* magazine, 1943. (G.S.)

Value $45-60

2.*Roy Rogers Souvenir Book*, 1940s. (G.S.)

Value $45-60

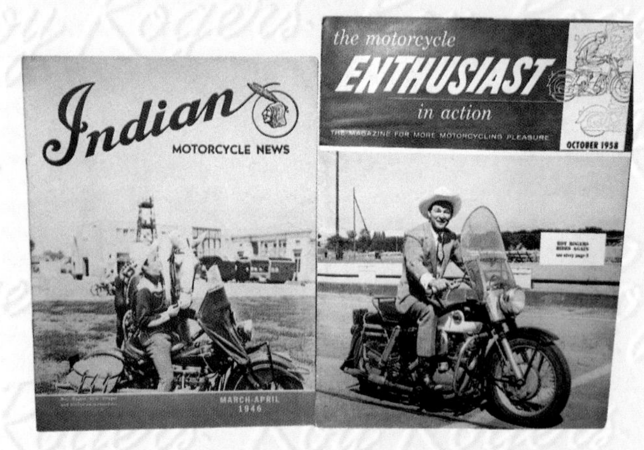

1. Roy Rogers on the cover of *Indian Motorcycle News*, 1946. (G.S.)
2. Roy Rogers on the cover of *Motorcycle Enthusiast* magazine, 1958. (G.S.)

Value $90-135 each

Roy Rogers on the cover of 1953 and 1954 issues of *Jet* magazine. (G.S.)

Value $45-90 each

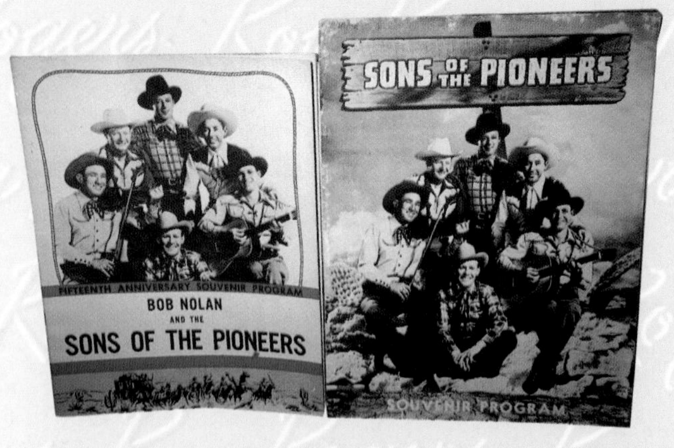

Bob Nolan and The Sons of the Pioneers souvenir programs, late 1930s, very rare. (G.S.)

Value $90-150 each

1. Roy Rogers on the cover of Thrill Circus Program. (G.S.)

Value $25-35

2.Roy Rogers and Dale Evans on the cover of Kentucky Fair program, 1958. (G.S.)

Value $25-40

1. Roy Rogers on the cover of *Western Family* magazine, 1943. (G.S.)
2. Roy Rogers on the cover of *Child Life* magazine, March 1943. (G.S.)

Value $45-75 each

1. Late 1930s photo of Roy on the cover of a Sunday newspaper. (G.S.)

Value $55-90

2. Late 1940s photograph of Dale Evans on the cover of a Sunday newspaper. (G.S.)

Value $45-80

Roy Rogers on the covers of *Western Stars* magazines produced by the Dell Company, 1950s. (R.L.)

Value $35-75 each

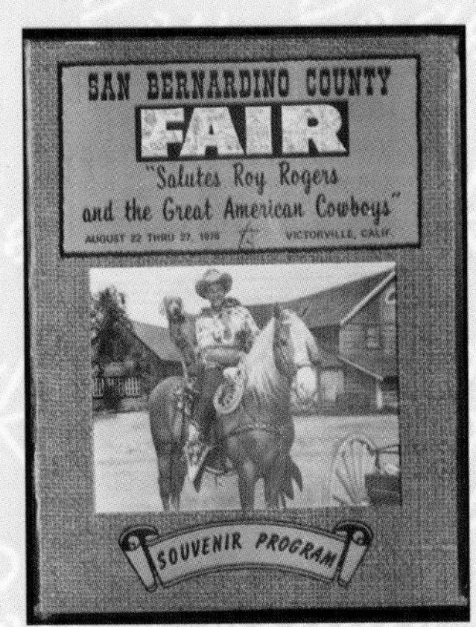

San Bernardino County Fair program with Roy Rogers on the cover, 1978. (C.Q.)

Value $15-30

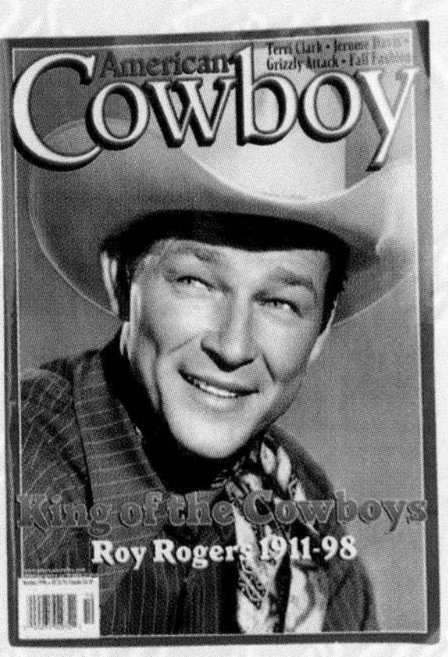

Roy Rogers on the cover of *American Cowboy* magazine, 1998. (R.L.)

Value $5-15

Roy Rogers and Dale Evans on the cover of *Possibilities* magazine, 1990s. (C.Q.)

Value $5-15

Roy Rogers and Dale Evans on the cover of *Hollywood Studio* magazine, 1980s. (C.Q.)

Value $20-35

CHAPTER 15

MISCELLANEOUS ITEMS

Roy Rogers Radio Show promotion folder 10″ x 12″ included 8″ x 10″ photograph of Roy and Trigger, rare. (G.S.)

Value $150-300

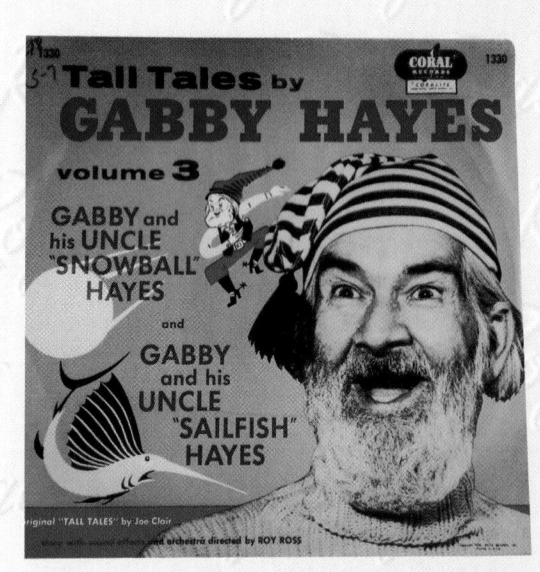

Gabby Hayes story-telling 45-rpm record produced by Coral Records. (M.M.)

Value $20-35

Roy Rogers gold-plated ashtray, 5″ diameter, raised relief design with Roy on rearing Trigger, 1950s. (C.Q.)

Value $50-90

Post Cereal's Roy Rogers Family Contest second-prize winner's certificate. Status that the winner had won a Bell and Howell movie camera, rare. (M.M.)

Value $75-125

Roy Rogers featured on Post's Grape-Nuts Flakes large and small cereal boxes. Rare unopened large box. (G.S.)

Value $75-150 small box

Value $250-350 large box

Quaker Oats Puffed Rice cereal box with advertisement for Gabby Hayes Cannon Ring, unopened box, extremely rare. (G.S.)

Value $250-350

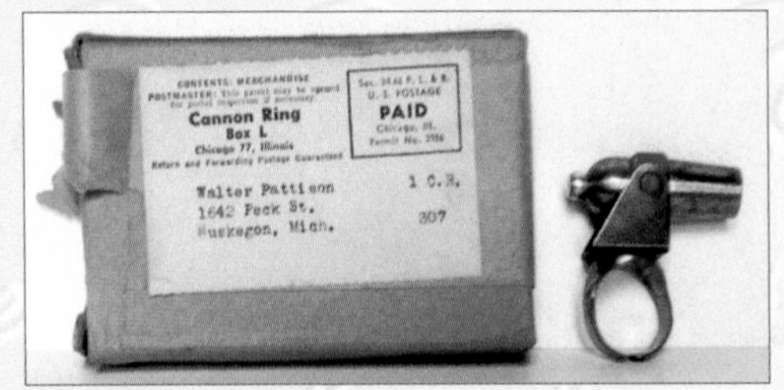

Gabby Hayes Cannon Ring with original box, extremely rare. (M.M.)

Value $250-350

1. Post's Wheat Meal cereal box with Roy Rogers coupon offer on back of box. Rare, unopened boxes. (G.S.)

Value $75-150

2. Roy Rogers pop/juice can, extremely rare. (G.S.)

Value $250-350

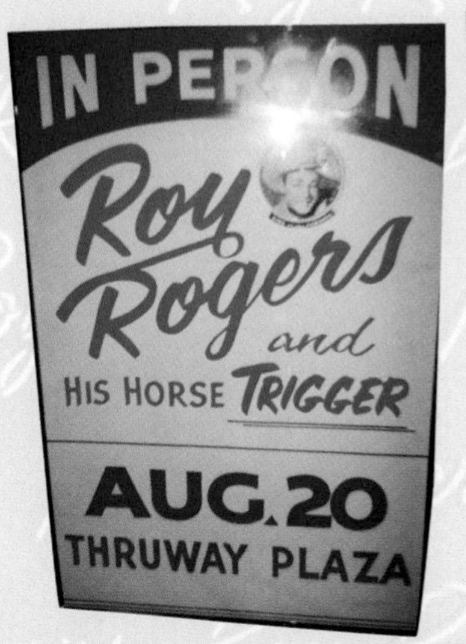

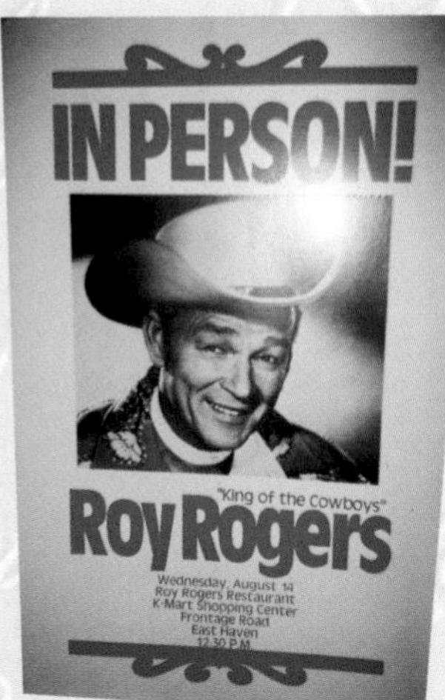

Large poster advertising Roy Rogers' personal appearances, rare. (G.S.)

Top row: Quaker Oats cereal boxes with Roy Rogers premium offers. (G.S.)

Value $125-250

Bottom row: Post's Wheat Meal cereal with Quaker Puffed Wheat Roy Rogers premium puzzle. (R.L.)

Value $75-150 Wheat Meal box

Value $15-30 Puzzle

Roy Rogers and rearing Trigger hairbrush. (D.T.)

Value $90-135

Roy Rogers black leather studded dog collar with red plastic jewels and Roy's name imprinted, rare. (M.M.)

Value $100-150

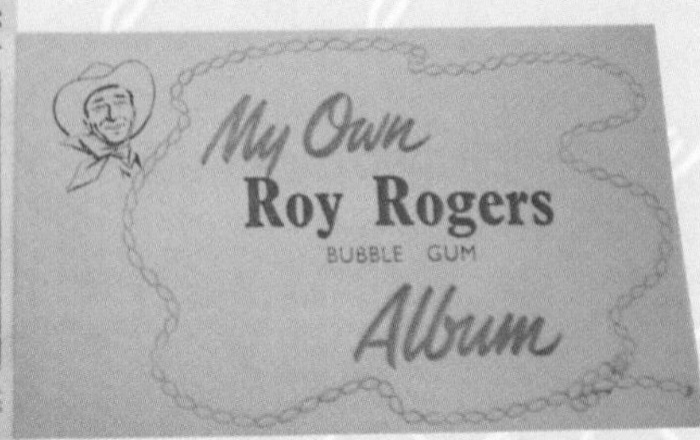

1. Order form for Roy Rogers cigarette cards. (M.M.)

Value $7-15

Roy Rogers Bubble Gum Album, 1950s, England. (M.M.)

Value $25-60

Roy Rogers Cookies box with unpunched spinner game and "Crackin' Good Gun" premiums, 1951–1953 by Carr Consolidated Biscuit Company, cookies made with Quaker Oats, honey, raisins, butter, flour, milk, and nuts. (G.S.)

Value $550-700

Gabby Hayes story-telling 45-rpm records produced by RCA Victor. "MacFadden and his Wonderful Lump" and "Allee Bamee and the Forty Horse Thieves." (M.M.)

Value $20-35 each

Gabby Hayes portable cook stove with instructions. (M.M.)

Value $125-250

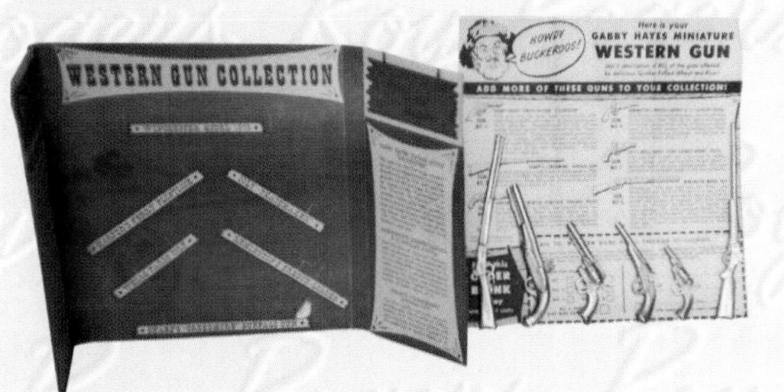

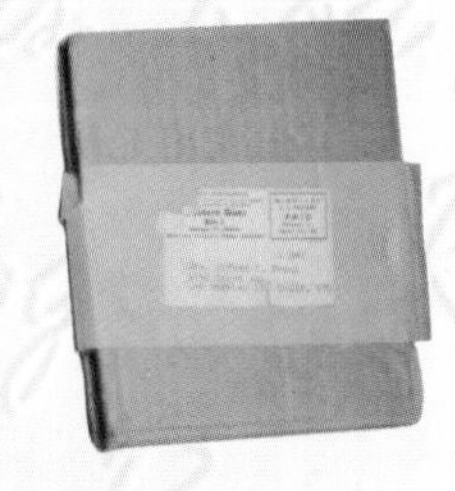

Gabby Hayes miniature western gun collection. Complete set with shipping box, extremely rare. (M.M.)

Value $350-500

Roy, Dale, and Trigger featured on Nestle's chocolate bar coupon. (C.Q.)

Value $7-10

Roy Rogers and Trigger pocket flashlight, 1950s. (M.M.)

Value $25-45

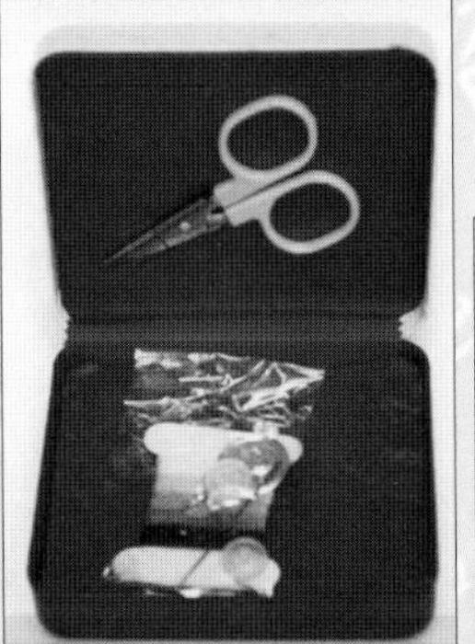

Dale Evans sewing kit, leatherette covered with Roy and Dale's brand logos on borders; 4" x 4" kit contains scissors, thread, needles, and thimble, 1950s, rare. (M.M.)

Value $65-110

Large printing photo-block negative, acid-etched metal, very rare. (G.S.)

Value Unknown

Gabby Hayes set of pictures in plastic Quaker Oats marked frames. Pictures and frames are premium items from Quaker Oats cereal. (M.M.)

Value $15-30 each

Printing photo-blocks for movies *Bells of Roserita*, *T.J. and Macintosh*, and *Yellow Rose of Texas*. Rare. (M.M.)

Value Unknown

CHAPTER 16

MOVIE POSTERS AND LOBBY CARDS

Leonard Slye was a member of "The Sons of the Pioneers" when he got his first acting job in the movie, *The Old Homestead* produced in 1935 by Liberty Pictures. Leonard, called Dick Weston by Liberty Pictures, played supporting roles in eleven more movies with stars like Gene Autry until 1938 when he began to star in his own movies.

Republic Pictures changed his name to Roy Rogers and gave him the starring role in the film, *Under Western Stars*. Roy had the talent and public appeal that made him a star in a very short time. He went on to make 90 more movies for Republic Pictures, Roy's last movie produced by Republic was *Pals of the Golden West*, in 1951. In 1952, Roy co-starred with Bob Hope and Jane Russell in Paramount's film *Son of Paleface*. The last 2 films in his active career were *Alias Jessie James*, produced by United Artists in 1959, and *Mackintosh and T.J.*, produced by Penland Productions in 1975.

Movie production companies created movie posters, lobby cards, and inserts (half-sheet posters) as advertisements for their films. One-sheet posters ranged in size from 22" x 28" to 40" x 60". They were printed on heavy stock paper using the lithograph process. Lithographed posters, inserts, and lobby cards for Republic Pictures films were printed by American Lithography Corp., Allied Printing Co., Essex Co., and Morgan Litho Co. The artwork was supplied by artists working for Republic Productions, Inc. Photographs (stills) taken of actual scenes from the western films were used mainly for the creation of lobby cards.

A set of eight lobby cards was produced and given to theaters for promotional purposes for each Roy Rogers movie. Each set contained one title card and seven scene cards from the film. The title card features the stars of the movies and is considered more valuable than the scenic cards. Lobby cards measure 11" x 14" and 12 1/2" x 16".

The wonderful illustrations created in brilliant colors by artists such as Peter Alvarado and John Usher make Roy Rogers movie posters, inserts, and lobby cards highly collectible and valuable as pieces of art and memorabilia. The only down side for a collector is having enough wall space to properly display them.

Generally, older posters, inserts, and lobby cards are worth more than newer ones. These items produced in the same year are worth about the same price. One-sheet movie posters are valued about four times as much as lobby cards from the same movie. Factors that determine value are age, condition, and rarity.

ROY ROGERS APPEARED IN TWELVE FILMS BEFORE HE STARRED IN HIS FIRST MOVIE *UNDER THE WESTERN STARS*, PRODUCED BY REPUBLIC PICTURES IN 1938.

The Old Homestead – Liberty, 1935.
Slightly Static – Short Subject, 1935.
Tumbling Tumbleweeds – Republic, 1935.
Way Up Thar – Short Subject, 1935.
Gallant Defender – Columbia – 1935.
The Mysterious Avenger – Columbia, 1936
Rhythm on the Range – Paramount, 1936.
The Big Show – Republic, 1936.
The Old Corral – Republic, 1936.
The Old Wyoming Trail – Columbia, 1937.
Wild Horse Rodeo – Republic, 1937.
The Old Barn Dance – Republic, 1938.

ROY ROGERS MOVIES PRODUCED BY REPUBLIC PICTURES, EXCEPT WHERE NOTED:

Alias Jesse James, United Artists, 1959
Along the Navajo Trail, 1945
Apache Rose, 1947
Arizona Kid, The, 1939
Arkansas Judge, 1941
Bad Man of Deadwood, 1941
Bells of Coronada, 1950
Bells of Rosarita, 1945
Bells of San Angelo, 1947
Billy the Kid Returns, 1938
Border Legion, The, 1940
Brazil, 1944
Carson City Kid, The, 1940
Colorado, 1940
Come On, Rangers, 1938
Cowboy and the Senorita, The, 1944
Dark Command, The, 1940
Days of Jesse James, 1939
Don't Fence Me In, 1945
Down Dakota Way, 1949
Eyes of Texas, 1948
Far Frontier, The, 1948
Frontier Pony Express, 1939
Gay Ranchero, The, 1948
Golden Stallion, The, 1949
Grand Canyon Trail, 1948
Hands Across the Border, 1943
Heart of the Golden West, 1942
Heart of the Rockies, 1951
Heldorado, 1946
Hit Parade of 1947, 1947
Hollywood Canteen, 1944
Home in Oklahoma, 1946
Idaho, 1943
In Old Amarillo, 1951
In Old Caliente, 1939
In Old Cheyenne, 1941
Jeepers Creepers, 1939
Jesse James at Bay, 1941
King of the Cowboys, 1943
Lake Placid Serenade, 1944
Lights of Old Santa Fe, 1944
Mackintosh and T.J., Penland Productions, 1975
Man From Cheyenne, 1942
Man From Music Mountain, The, 1943
Man From Oklahoma, The, 1945
Melody Time, 1948
My Pal Trigger, 1945
Nevada City, 1941
Night Time in Nevada, 1948
North of the Great Divide, 1950
On the Old Spanish Trail, 1947
Out California Way, 1946
Pals of the Golden West, 1951
Rainbow Over Texas, 1945
Ranger and the Lady, The, 1940
Red River Valley, 1941
Ridin' Down the Canyon, 1942
Robin Hood of the Pecos, 1941
Roll On, Texas Moon, 1946
Romance on the Range, 1942
Rough Riders' Roundup, 1938
Saga of Death Valley, 1939
San Fernando Valley, 1944
Sheriff of Tombstone, 1941
Shine on, Harvest Moon, 1938
Silver Spurs, 1943
Song of Arizona, 1945
Song of Nevada, 1944
Song of Texas, 1943
Son of Paleface, Paramount, 1952
Sons of the Pioneers, 1941
South of Caliente, 1951
South of Santa Fe, 1942
Southward Ho, 1939
Spoilers of the Plains, 1951
Springtime in the Sierras, 1947
Sunset in El Dorado, 1945
Sunset in the West, 1950
Sunset on the Desert, 1942
Sunset Serenade, 1942
Susanna Pass, 1949
Trail of Robin Hood, 1950
Trigger, Jr., 1950
Twilight in the Sierras, 1950
Under California Stars, 1948
Under Nevada Skies, 1946
Under Western Stars, 1938
Utah, 1945
Wall Street Cowboy, 1939
Yellow Rose of Texas, The, 1944
Young Bill Hickok, 1940
Young Buffalo Bill, 1940

Heart of the Golden West. Republic Pictures, 1942. (G.S.)
Value $225-350

Pals of the Golden West. Republic Pictures, 1951. (G.S.)
Value $160-280

Wall Street Cowboy. Republic Pictures, very rare, 1939. (G.S.)
Value $250-400

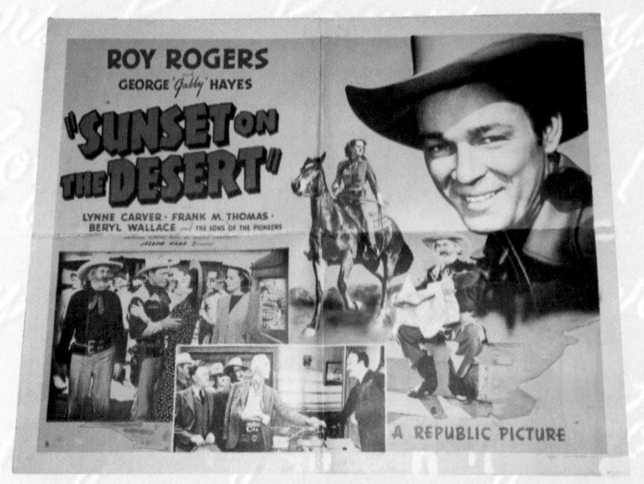

Sunset on the Desert. Republic Pictures, 1942. (G.S.)
Value $225-350

South of Santa Fe. Republic Pictures, 1942. (G.S.)
Value $225-350

Bad Man of Deadwood. Republic Pictures, 1942. (G.S.)
Value $225-350

On The Old Spanish Trail. Republic Pictures, 1947. (G.S.)
Value $210-325

Pals of The Golden West. Banner. Republic Pictures, 1951. (G.S.)
Value $175-250

Down Dakota Way. Republic Pictures, 1949. (G.S.)
Value $180-300

The Far Frontier. Republic Pictures, 1948. (G.S.)
Value $180-300

Eyes of Texas. Insert. Republic Pictures, 1949. (G.S.)
Value $200-350

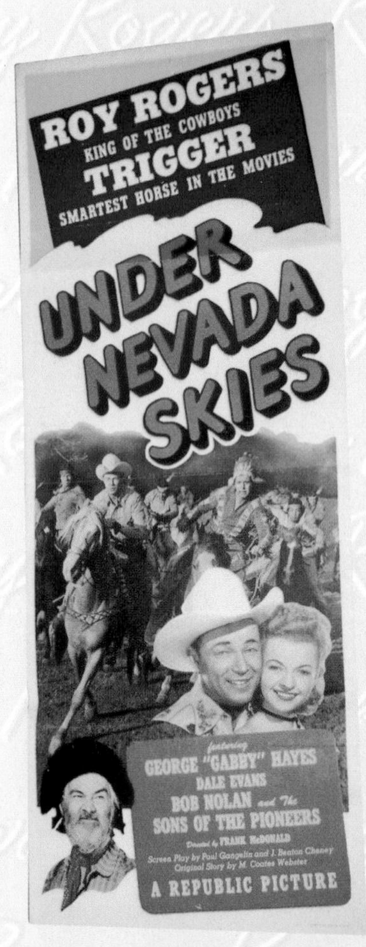

Under Nevada Skies. Insert 14" x 36". Republic Pictures, 1946. (G.S.)
Value $250-350

Roll on Texas Moon. Insert. 14″ x 36″. Republic Pictures, 1946. (G.S.)
Value $225-350

My Pal Trigger. Insert. 14″ x 36″. Republic Pictures, 1946. (G.S.)
Value $300-450

Heart of The Rockies. Insert. 14″ x 36″. Republic Pictures, 1951. (G.S.)
Value $175-275

The Golden Stallion. Insert. 14″ x 36″. Republic Pictures, 1949. (G.S.)
Value $225-350

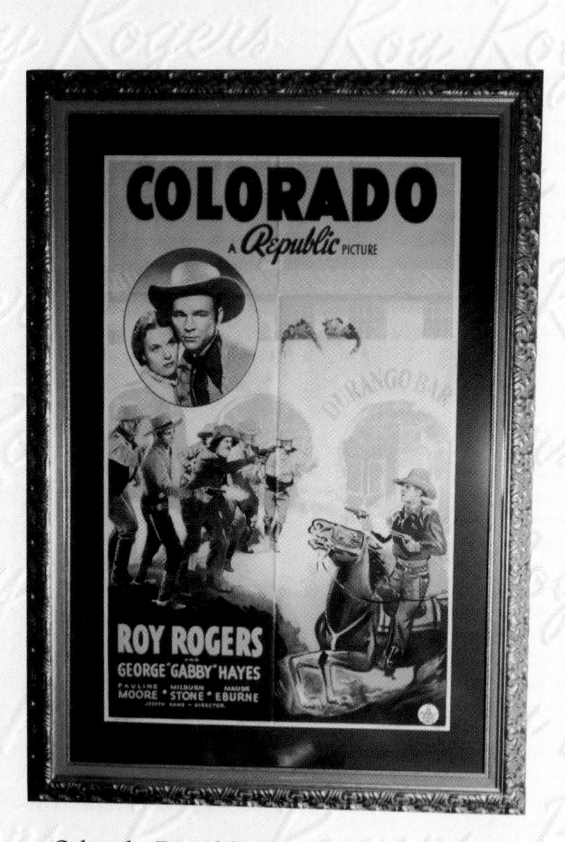

Colorado. Republic Pictures, 1940. (G.S.)
Value $240-375

The Old Corral. Republic Pictures, 1936. Starred Gene Autry. Rare lobby card with Roy and Gene shown together. (G.S.)
Value $75-150

Song of Arizona. Republic Pictures, 1945. (G.S.)
Value $35-75

Top:
1. *Casanova Burlesque*. Featuring Dale Evans. Republic Pictures, early 1940s. (G.S.)
Bottom:
2. *The Big Show-Off*. Featuring Dale Evans. Republic Pictures, early 1940s. (R.L.)
Value $30-60 each

Man From Oklahoma. Scene Cards.
Republic Pictures, 1945. (G.S.)
Value $35-70 each

1. *Song of Nevada*. Republic Pictures, 1943. (G.S.)
2. *Hands Across the Border*. Republic Pictures, 1943. (G.S.)
3. *Silver Spurs*. Republic Pictures, 1943. (G.S.)
4. *Man From Music Mountain*. Republic Pictures, 1943. (G.S.)

Value $35-70 each

Man From Music Mountain.
Republic Pictures, 1943. (G.S.)
Value $35-70

1. *The Ranger and the Lady*. Republic Pictures, 1941. (G.S.)
2. *Jessie James at Bay*. Republic Pictures, 1940. (G.S.)
3. *Man From Cheyenne*. Republic Pictures, 1946. (G.S.)
4. *Home in Oklahoma*. Republic Pictures, 1942. (G.S.)

Value $35-70 each

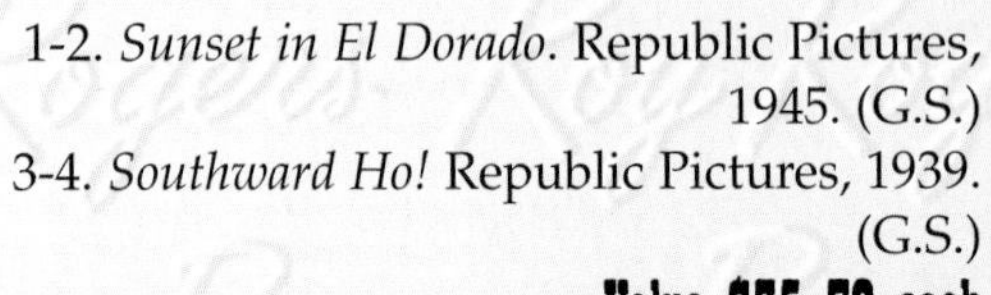

1-2. *Sunset in El Dorado*. Republic Pictures, 1945. (G.S.)
3-4. *Southward Ho!* Republic Pictures, 1939. (G.S.)

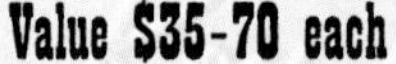

Value $35-70 each

1 *The Arizona Kid*. Republic Pictures, 1939. (G.S.)

Value $35-70

2. *In Old Caliente*. Republic Pictures, 1939. (G.S.)

Value $35-70

3. *Shine on Harvest Moon*. Title Card. Republic Pictures, 1938. (G.S.)

Value $60-110

4. *Sons of the Pioneers*. Republic Pictures, 1942. (G.S.)

Value $35-70

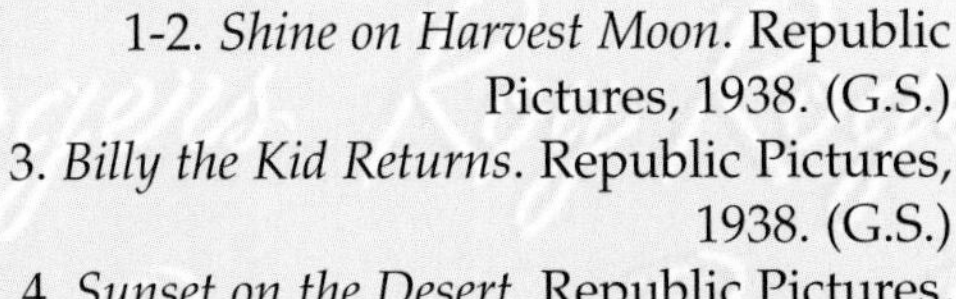

1-2. *Shine on Harvest Moon*. Republic Pictures, 1938. (G.S.)
3. *Billy the Kid Returns*. Republic Pictures, 1938. (G.S.)
4. *Sunset on the Desert*. Republic Pictures, 1938. (G.S.)

Value $35-70 each

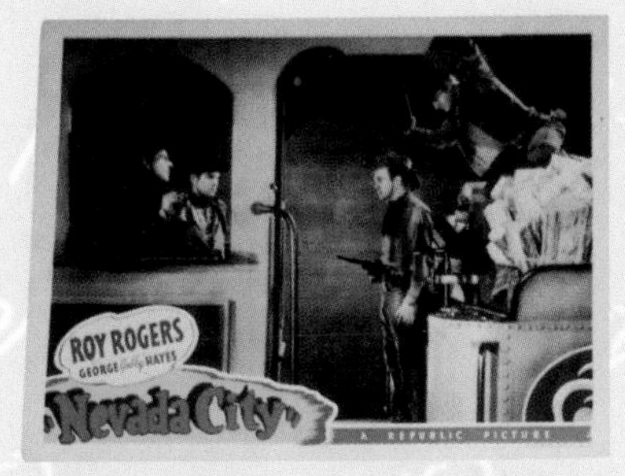

1-2. *Nevada City*. Republic Pictures, 1941. (G.S.)
3-4. *The Border Legion*. Republic Pictures, 1940. (G.S.)

Value $35-70 each

1. *Roll on Texas Moon*. Title Card. Republic Pictures, 1946. (G.S.)
2. *Sunset on the Desert*. Title Card. Republic Pictures, 1942. (G.S.)
3. *Under California Stars*. Title Card. Republic Pictures, 1948. (G.S.)
4. *Song of Nevada*. Title Card. Republic Pictures, 1944. (G.S.)

Value $60-110 each

1. *Song of Texas*. Title Card. Republic Pictures, 1943. (G.S.)
2. *San Fernando Valley*. Title Card. Republic Pictures, 1943. (G.S.)
3. *Romance on the Range*. Title Card. Republic Pictures, 1942. (G.S.)
4. *Sunset Serenade*. Title Card. Republic Pictures, 1942. (G.S.)

Value $60-110 each

San Fernando Valley. Scene cards. Republic Pictures, 1943. (G.S.)

Value $35-70 each

1. *The Golden Stallion*. Title Card. Republic Pictures, 1949. (G.S.)
2. *Song of Arizona*. Title Card. Republic Pictures, 1945. (G.S.)
3. *Eyes of Texas*. Title Card. Republic Pictures, 1948. (G.S.)
4. *Man From Oklahoma*. Title Card. Republic Pictures, 1945. (G.S.)

Value $60-110 each

1. *Susanna Pass*. Title Card. Republic Pictures, 1949. (G.S.)
Value $60-110

2. *Down Dakota Way*. Title Card. Republic Pictures, 1949. (G.S.)
Value $60-110

3. *Robin Hood of the Pecos*. Title Card. Republic Pictures, 1941. (G.S.)
Value $60-110

4. *Down Dakota Way*. Republic Pictures, 1949. (G.S.)
Value $35-70

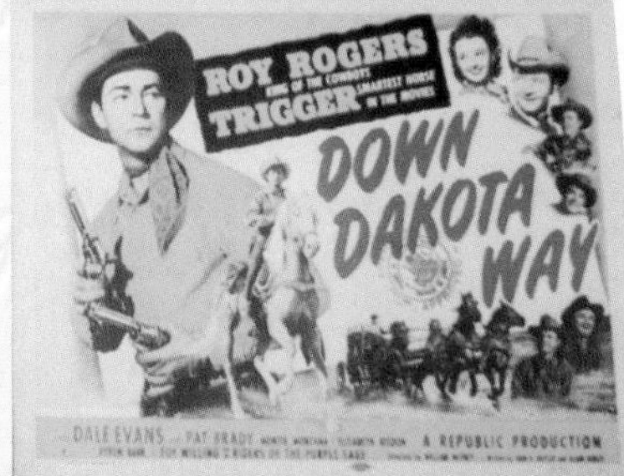

Heart of The Golden West. Title card. Republic Pictures, 1942. (G.S.)
Value $75-150

King of the Cowboys. Title card. Republic Pictures, 1943. (G.S.)
Value $65-140

1. *Under California Stars*. Title Card. Republic Pictures, 1948. (G.S.)
2. *Night Time in Nevada*. Title Card. Republic Pictures, 1948. (G.S.)
3. *The Gay Ranchero*. Title Card. Republic Pictures, 1948. (G.S.)

Value $60-110 each

1. *Rainbow Over Texas*. Title Card. Republic Pictures, 1945. (G.S.)
2. *South of Caliente*. Title Card. Republic Pictures, 1951. (G.S.)
3. *In Old Amarillo*. Title Card. Republic Pictures, 1951. (G.S.)
4. *North of the Great Divide*. Title Card. Republic Pictures, 1950. (G.S.)

Value $60-110 each

Sunset in The West. Title Card. Republic Pictures, 1950. (G.S.)

Value $50-110

Trail of Robin Hood. Autographed Title Card. Republic Pictures, 1950. (C.Q.)

Value $50-110. Add $50 for autograph.

1. *Heart of the Golden West*. Title Card. Republic Pictures, 1949. (G.S.)
2. *Susanna Pass*. Title Card. Republic Pictures, 1949. (G.S.)
3. *Pals of the Golden West*. Republic Pictures, 1951. (G.S.)
4. *Heart of the Rockies*. Republic Pictures, 1951. (G.S.)

Value $60-110 each

1. *Melody Time*. Republic Pictures, 1948. (G.S.)
Value $25-40

2. *Eyes of Texas*. Republic Pictures, 1948. (G.S.)
Value $35-70

3. *Springtime in the Sierras*. Republic Pictures, 1947. (G.S.)
Value $35-70

4. *Rainbow Over Texas*. Republic Pictures, 1946. (G.S.)
Value $35-70

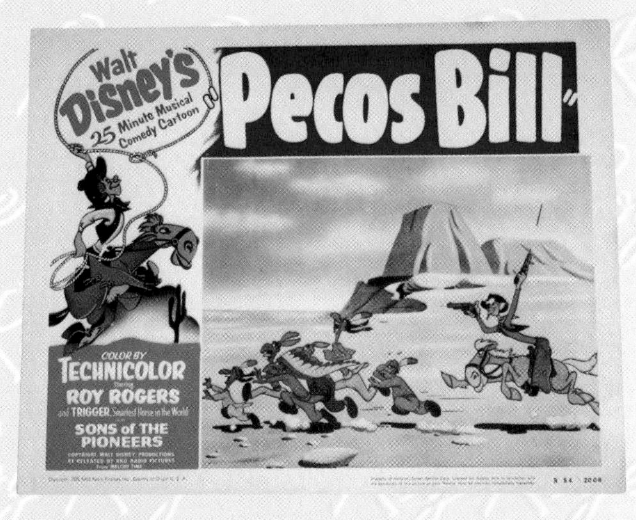

Pecos Bill. Walt Disney Productions, 1953. (G.S.)
Value $25-60

Melody Time. Republic Pictures, 1948. Distributed by RKO Radio Pictures, Mexico. (G.S.)
Value $25-60

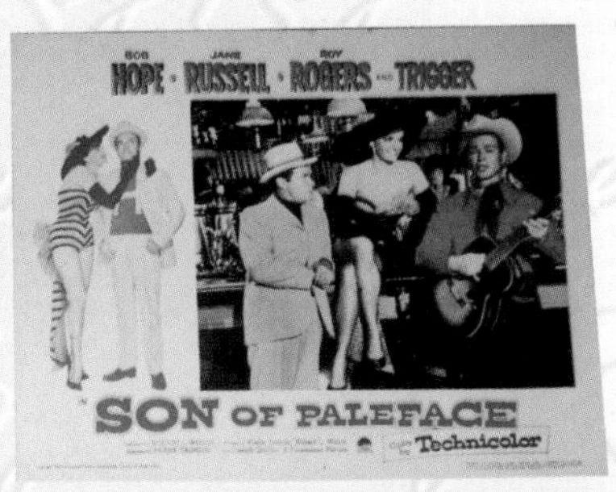

1.-3. *Son of Paleface*. Paramount, 1952. (G.S.)
Value $35-70 each

4. *Brazil*. Republic Pictures, 1944. (G.S.)
Value $35-50

CHAPTER 17

MUSIC

Roy Rogers and Dale Evans were professional singers many years before they became actors. Roy's parents taught him to play guitar, yodel, and sing at a very early age. By the time he was twelve, Roy was playing guitar, yodeling, and singing at local dances. People loved to hear him sing their favorite cowboy songs. His first job as a singer and guitarist was with a western group called the "Rocky Mountaineers." In a short time, Roy left this group and started his own "O–Bar–O" band. After an unsuccessful tour of the Southwest, the group disbanded.

In 1933, Tim Spencer, Bob Nolan, and Roy formed "The Pioneer Trio" and performed on radio station KFWB in Hollywood. A year later, Hugh Farr joined the trio and the group's name was changed to "The Sons of the Pioneers." Their first big break was a recording contract with Decca Records. They recorded several hundred songs, and their song, "Tumbling Tumbleweed" reached number 13 on the charts in 1934. The group's first songbook was published in 1939 and titled, *"The Sons of the Pioneers Song Folio #1."*

The previous year, Roy got a contract with Republic Pictures to make western movies. In his acting career, Roy made 88 musical western movies, and was voted the number one singing cowboy in the mid-1940s. He recorded over 350 songs in his lifetime.

Roy Rogers Cowboy Band set with two-piece box by Spec-Toy-Culars, Inc., box dimensions 17 1/2" x 14 1/2" x 3", multi-colored plastic instruments, 1953, rare. (G.S.)

Value $350-600

Roy Rogers Western Band set by Emenee Company, two-piece box, 1950s. (L.C.)
Value $350-450

Roy Rogers Cowboy Band set with song sheet and two-piece box. Gold-colored plastic instruments. Spec-Toy-Culars, Inc., early 1950s, 17 1/2″ x 14 1/2″ x 3″ rare. (G.S.)
Value $350-600

Roy Rogers wood guitar, sold by Sears, Roebuck & Company in two sizes, 36″ x 13″ and 33″ x 11″. Cowboy and horse campfire scene stenciled in brown and yellow, Roy's signature on handstock and on bottom, 1950s. (C.Q.)
Value $300-450

Roy Rogers full-sized 36″ x 13″ wood guitar with birch-wood top, and black-stenciled campfire scene with a horse and cowboy. Roy's facsimile signature on bottom of scene. (C.Q.)
Value $350-500

Roy Rogers full-sized 36″ x 13″ black wood guitar, tan and white stenciled campfire scene with a horse and cowboy and Roy's facsimile signature on bottom of scene, very rare. (C.Q.)
Value $400-550

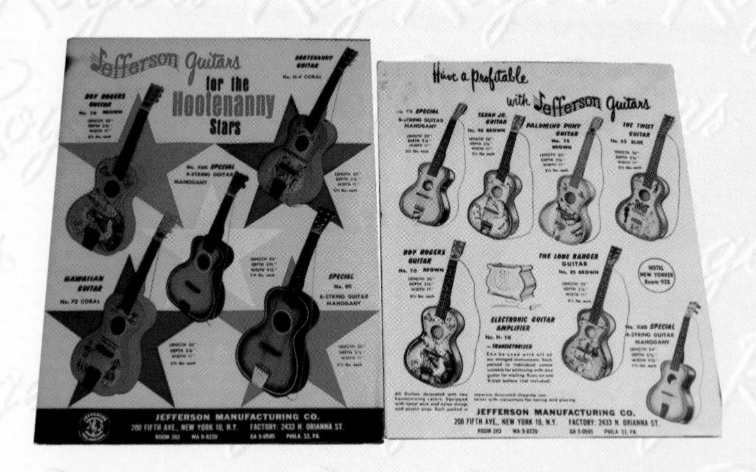

Jefferson Manufacturing Company, Inc. advertisements for children's toy guitars, including the Roy Rogers guitar made in the 1960s. The company is located in Philadelphia, Pennsylvania and is still going strong with George Jefferson at the helm.

Roy Rogers child's wood guitar with original guitar case. Roy's facsimile signature stamped under sound hole inside of 33″ x 11″ guitar. Leatherette case has printed cowboy scenes, very rare. (C.Q.)

Value $250-350, add $100 for case

Roy Rogers red toy guitar with white stenciled scene of Roy on rearing Trigger, measures 30″ x 11″ made by Range Rhythm Toys, Inc., 1950s. (C.Q.)

Value $100-200

Roy Rogers toy red guitar 30″ x 11″ with white stenciled images of Roy and Trigger, 1960s. (C.Q.)

Value $100-200

Roy Rogers child's toy guitar maker unknown, 30″ x 11″ yellow painted body with small cartoon-like cowboy stencils. Roy's facsimile signature at base of guitar, rare. (C.Q.)

Value $125-225

Roy Rogers children's guitar 30″ x 11″ x 3 3/4″ by Jefferson Mfg. Co., Inc., stenciled image of Roy riding Trigger with Roy and Dale's portraits on upper area of guitar. (C.Q.)

Value $100-200, add $50 for shipping box

Roy Rogers Limited Edition Toy Guitar made by Jefferson Mfg. Co., Inc. This guitar was recently made to honor the life and legacy of Roy Rogers. (R.L.)

Value $30-45, new in sealed display box

Roy Rogers children's guitar, measures 30″ x 11″ x 3 3/4″ by Jefferson Mfg. Co., Inc. Stenciled images of Roy, Dale, Trigger, and Roy Rogers brand logo, painted orange and trimmed in white, tan, and cream colors. Roy's facsimile signature on bottom of guitar, 1960s. (C.Q.)

Value $100-200

Roy Rogers Riders Harmonica on sealed display card, engraved metal 4 1/2″ long, 1950s. (C.Q.)

Value $85-125

Roy Rogers Riders Harmonica and Roy Rogers Cowboy Band Harmonica. The Cowboy Band Harmonica is the rarest. Only 2000 were made and given as prizes in a Roy Rogers radio contest sponsored by Quaker Oats. Precision-tuned with a solid brass plate. (C.Q.)

Value $110-175 with box

Roy Rogers and Dale Evans record player, silhouettes of Roy and Dale on detachable speakers, image of Roy on rearing Trigger on the main body inside lid, 1959. (C.Q.)

Value $350-550

Roy Rogers Happy Trails 45-rpm record player. Illustrations of cowboy scenes and pistols on hard plastic with cream-colored case. Made by RCA Corp. in the 1950s. (Roy Rogers Museum)

Value $250-400

RECORD ALBUMS

Top row: *"Happy Trails"* with Roy Rogers and Dale Evans signing "The Yellow Rose of Texas" 45- and 78-rpm records, hi fidelity, RCA Victor Records. (G.S.)

Value $60-125 each

Bottom row: *"Dale Evans Sings"* 10" hi fidelity 45- and 78-rpm records, 1940s. (G.S.)

Value $45-115 each

Top row: "The Masked Marauder" featuring Roy Rogers, 45- and 78-rpm records by RCA Victor, early 1950s. (G.S.)

Value $60-125 each

Bottom row: *Hymns of Faith Songs* by Roy Rogers and Dale Evans, RCA Victor Records, two-record set, 1954. (G.S.)

Value $45-115

Roy Rogers premium record from Post's Sugar Crisp Cereal, rare, 1950s. (M.M.)

Value $45-75

Roy Rogers and Dale Evans 45-rpm record with jacket, "The Old Chisholm Trail" and "Red River Valley" by Little Golden Records, produced by the Sandpiper Press, 1950s. (Roy Rogers Museum)

Value $20-35 each

Top row: Roy Rogers Rodeo Albums, 45- and 78-rpm, recorded by RCA Victor Records, 1950s. (G.S.)

Value $60-125 each

Bottom row: Roy Rogers Souvenir Albums, two record sets in 45- and 78-rpm by RCA Victor Records, 1952. (R.L.)

Value $60-125 each

Cowboy Classics sung by The Sons of the Pioneers record sets, RCA Victor Records. (G.S.)

Value $45-125 each

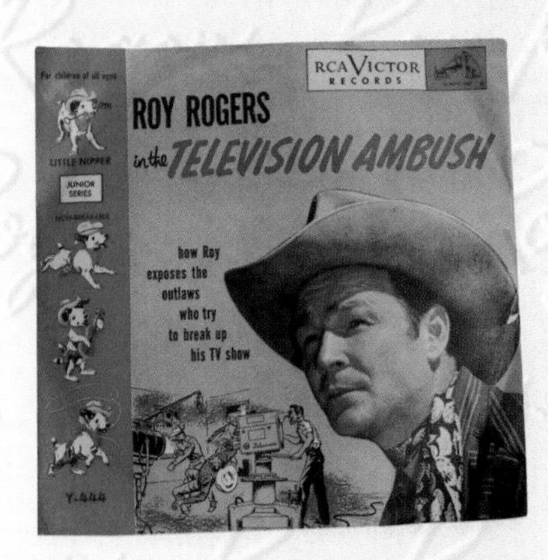

Roy Rogers in "The Television Ambush" 45-rpm record by RCA Victor Records, 1950s. (Roy Rogers Museum)

Value $20-35

Roy Rogers and Dale Evans 78-rpm record album *The West* includes "*A Child's Introduction to the West*" and "*16 Great Songs of the West*", includes records, pictures, and stories, 1950s. (G.S.)

Value $75-150 each

1. *Lore of the West in Song and Story* with Roy Rogers and Gabby Hayes, RCA Victor Records, a two-record set, the *Little Nipper* series, 1950s. (G.S.)
2. *Rip Roaring Adventures of Pecos Bill*, songs by Roy Rogers and Sons of the Pioneers, RCA Victor Records. (G.S.)

Value $75-150 each

Roy Rogers Tells and Sings About Pecos Bill by RCA Victor Records. *Little Nipper* series 1950s album contains two 78-rpm records with pictures and stories. (C.Q.)

Value $75-150

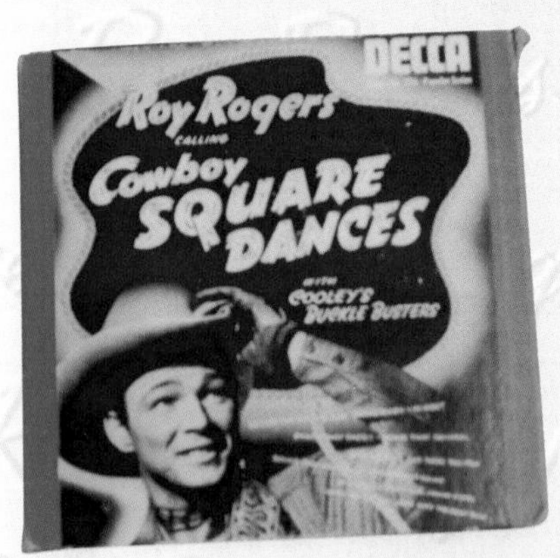

Roy Rogers Calling Cowboy Square Dances, Decca Records, early 1940s. (Roy Rogers Museum)

Value $60-125

Roy Rogers and Dale Evans Sing "Open up your Heart" and "The Lord is Counting on You." Golden Records 45-rpm. (R.L.)

Value $15-20

"*Jesus Loves Me*" and other songs sung by Roy Rogers and Dale Evans, on 78-rpm, Pickwick and RCA, 1960s. (R.L.)

Value $15-20

Roy's photo on cover of *Songs of the Soil* song folio, 1948. (R.L.)
Value $35-60

Roy Rogers on cover of *Popular Western Music Folio of Hit Songs*, 1940s. (R.L.)
Value $30-45

"When My Blue Moon Turns to Gold Again," Roy Rogers on cover, 1940s. (C.Q.)
Value $35-60

Songs of the Soil, Roy's picture in black and white on cover and full-colored picture on back of songbook, 1948. (G.S.)
Value $35-65

The Bible Tells Me So songbook, featured by Roy Rogers and Dale Evans, words and music by Dale Evans. Paramount-Roy Rogers Music Company, 1954. (R.L.)
Value $35-60

The Sons of the Pioneers song folios #3 produced in 1943 by American Music, Inc., Folios #1 and #2 not shown. (R.L.)
Value $35-60 each

"I Dream of Jeannie" by Stephen Foster, photo of Dale Evans on cover, 1940s, autographed by Dale Evans. (C.Q.)
Value $25-50, without autograph, add $35 with autograph

Song Hits magazine with Roy Rogers on cover, 1940s (C.Q.)
Value $35-60

Roy Rogers Song Folio by Paramount–Roy Rogers Music Company, Inc., 1952. (C.Q.)
Value $25-50

Roy Rogers Guitar Folio by Paramount–Roy Rogers Music Co., Inc., 1952. (C.Q.)
Value $25-50

Top Hits–Western Songs with Roy Rogers on Cover, 1940s. (C.Q.)
Value $45-75

Roy Rogers Favorite Cowboy Songs, by Robbins Music, 72 pages, 1943. (R.L.)
Value $45-75

Roy Rogers Album of Cowboy Songs by Edwards and Marks Music Co., 1943. (C.Q.)
Value $35-70

"California Rose" by Famous Music Co. Words and Music from Paramount Films *Son of Paleface* with Roy Rogers and Bob Hope, 1950s. (C.Q.)

Value $12-20

"Cattle Call" words and music by Tex Owens, featured by Roy Rogers, 1940s. (C.Q.)

Value $35-70

Roy Rogers Song Folio by Paramount–Roy Rogers Music Co., Inc. (G.S.)

Value $25-50

Roy Rogers on songbook covers:
1. *Daddy's Little Cowboy* (G.S.)
2. *Hazy Mountains* (G.S.)
3. *Merry Christmas, My Darling* (G.S.)
4. *Hawaiian Cowboy* (G.S.)

Value $25-50 each

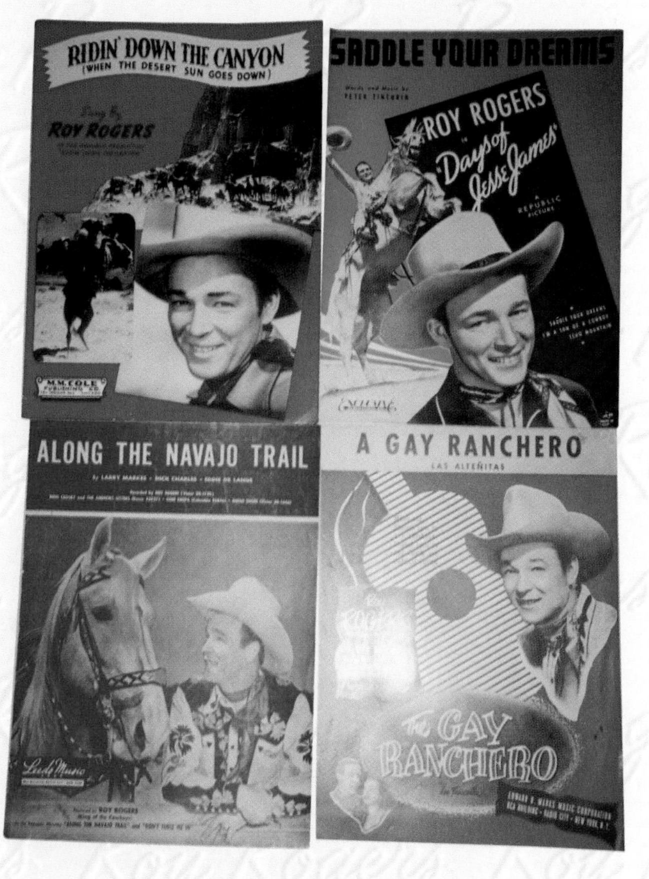

Roy Rogers on songbook covers:
1. *Ridin' Down the Canyon* (G.S.)
2. *Saddle Your Dreams* (G.S.)
3. *Along the Nevaja Trail* (G.S.)
4. *The Gay Ranchero* (G.S.)

Value $25-50 each

1. *Roy Rogers Own Songs*, American Music, Inc., 1943. (G.S.)
2. *Jamboree Songbook* with Roy Rogers on the cover, 1940s. (R.L.)

Value $30-60 each

1. Dale Evans on cover of "Corral in the Sky" song folio 1940s. (G.S.)
2. Dale Evans on cover of "The Sunshine of Paradise Alley" early 1940s. (G.S.)

Value $15-25 each

1. "Hands Across the Border" featured by Roy Rogers and the Sons of the Pioneers, 1940s. (G.S.)
2. "These Tears Are Not For You" with Roy Rogers, 1940s. (G.S.)
3. "You Should Know" song sheet featured Roy Rogers, 1940s. (G.S.)

Value $25-50 each

Roy Rogers featured on early 1940s song sheets. (G.S.)
1."Man from Music Mountain"
2."There's a Cloud in my Valley of Sunshine"
3."On the Old Spanish Trail"
4."I'm Gonna' Have a Cowboy Weddin'"

Value $25-50 each

CHAPTER 18

PICTURES OF ROY ROGERS AND DALE EVANS

There were literally thousands of publicity photos taken of Roy Rogers and Dale Evans from the 1930s to the 1990s. These photographs were used in all types of media including music folios, books, movie and event promotions, magazine and newspaper articles, fan club letters, comics, coloring and activity books, and general merchandizing.

Republic Pictures had countless photos taken of Roy and Dale to promote their western movies. Many were actual action scene shots called "stills." Other publicity still photographs were taken after the completion of a movie in a studio with cast members. These were usually printed in 8" x 10" size and the originals will have a number and stamped name of the studio on them, or the company that used them. Most of these original publicity still photographs of Roy and Dale have been reproduced many times over the years. Facsimile signatures of Roy and Dale have been imprinted on some of these photographs. Reproduced photographs generally have a value of $3–5.

Candid photographs taken of Roy Rogers and Dale Evans between 1911 and the late 1930s are extremely rare and belong mainly to family members. There are, however, thousands of original candid photographs of Roy and Dale taken at personal appearances and various events from the 1940s through the 1990s that remain. These candid photographs usually stay with the person who took them, but at times will be offered for sale. Candid photographs of Roy and Dale are valued from $15–100 and earlier photographs are generally worth more.

Between 1940 and 1950, Dixie Ice Cream Company issued 8" x 10" color premium pictures of Roy Rogers, Dale Evans, Gabby Hayes, Pat Brady, Jingles, and Trigger. These pictures were publicity photographs taken by Republic Pictures of various Roy Rogers and Dale Evans movies that were made in this time period. Value of these pictures has greatly increased in the past few years as more people are collecting them.

Lids of Dixie ice cream cups advertising free movie stars pictures. (C.Q.) Reverse side of Dixie ice cream cup lids with Republic Pictures photos of Roy Rogers and Gabby Hayes. (C.Q.)

Value $3-5 each

DIXIE PREMIUM PICTURES

Gabby Hayes Dixie premium 8″ x 10″ color picture with Republic Pictures name on border, late 1940s. (C.Q.)

Value $45-90

Roy Rogers Dixie premium 8″ x 10″ color pictures with Republic Pictures name on border, 1940s. (C.Q.)

Value $45-90 each

Dale Evans Dixie premium 8″ x 10″ color picture with Republic Pictures name on border, 1940s. (C.Q.)

Value $45-90

Dixie premium pictures of Roy Rogers and co-stars Gabby Hayes, Andy Devine, Jane Frazer, and Penny Edwards. (G.S.)

Value $45-90 each

PRINTS

Limited edition print titled "Is She Thinking of Me or Tom Mix" by Gina Faulk. (C.Q.)
Value $375-550 signed
$75 unsigned

Limited edition print titled, "Saturday Morning TV Anywhere USA" by Sean Sullivan. Recent release. (C.Q.)
Value $375-550 signed
$75 unsigned

Roy Rogers "King of the Cowboys" illustrated black and white print by Manning Hall, 1950s. (Roy Rogers Museum)
Value Unknown

PHOTOGRAPHS

Roy Rogers and Trigger color photos, 1950s by Republic Pictures. (R.L.)
Value $5-10 each

Original 1950s candid photograph of Roy Rogers waving his hat. (C.Q.)
Value $75-100

Roy Rogers and Dale Evans color photos by Republic Pictures. Photos are shown on backs of book covers. (C.Q.)
Value $3-5 each

Black and white publicity photographs of Roy Rogers and Roy Rogers and Trigger, 8″ x 10″ size, by Republic Pictures, early 1950s. (C.Q.)
Value $3-10 each

Roy Rogers publicity photos used on the back of book covers. (C.Q.)
Value $3-5 each

Dale Evans and Buttermilk 8″ x 10″ black and white photograph. (C.Q.)
Value $6-12

Dale Evans publicity posed black and white photo by Republic Pictures. (C.Q.)
Value $6-12

Roy Rogers and Dale Evans 8″ x 10″ black and white posed publicity photograph by Republic Pictures, early 1950s. (R.L.)
Value $3-10

Dale Evans posed black and white 8″ x 10″ photograph, 1940s, by Republic Pictures. (C.Q.)
Value $10-15

Dale Evans 8″ x 10″ color photo with Buttermilk by Republic Pictures, 1950s still. (R.L.)
Value $6-12

CHAPTER 19

School Supplies and Related Items

Roy Rogers and Dale Evans dearly loved children. They showed their love by adopting six orphaned children and giving their time, energy, and resources to youth organizations, children's hospitals, and schools. They improved the welfare of young people by working with 4-H clubs, giving free performances at children's hospitals, and spearheading a safety program for school children. Awards were presented annually to schools that had the best safety record. One of the awards that Roy and Dale presented to a winning school was a gold-plated cast metal statuette of Trigger mounted on a walnut base. Between 1948 and 1960, thousands of schools participated in the Roy Rogers and Dale Evans National Safety Program.

Dale Evans red small schoolbag with plastic shoulder straps 11" x 8" 1950s plaid cotton material with plastic flap. (C.Q.)

Value $65-110

Original photo of Roy Rogers and Trigger on a simulated leather school bag, 1950s. (Roy Rogers Museum)

Roy Rogers, Dale Evans, and Trigger large 10″ x 14″ simulated leather school bag, late 1950s and early 1960s with photo on front. (G.S.)

Value $125-250

Roy Rogers and Trigger 12″ x 12″ simulated leather school bag with photograph of Roy riding Trigger. (Roy Rogers Museum)

Value $125-250

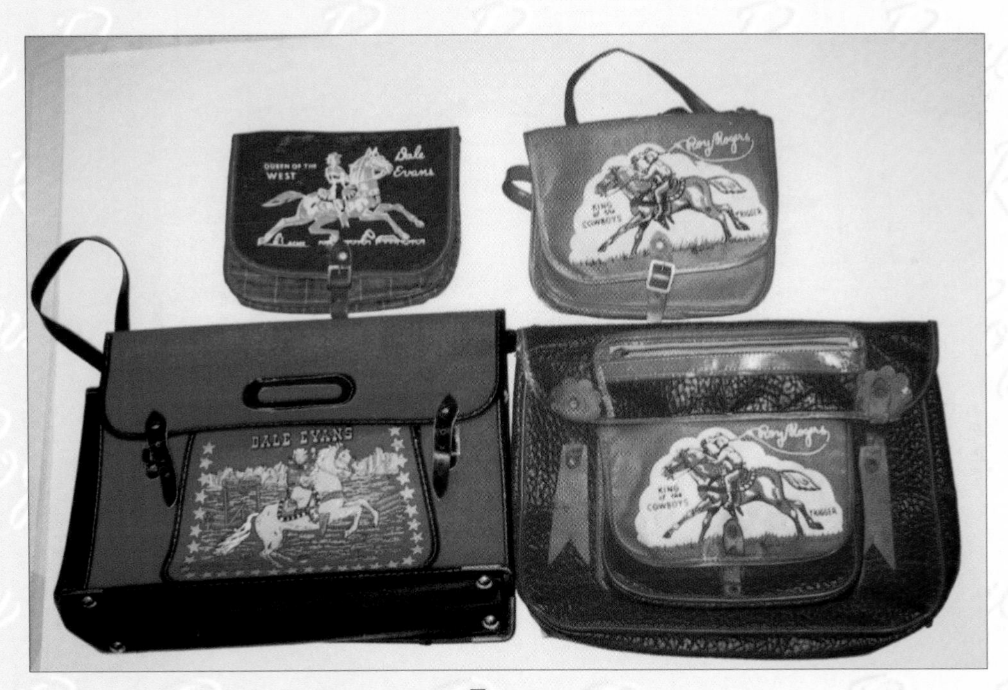

Top row:

1. Dale Evans small school bag 11″ x 8″ illustrated image of Dale on Buttermilk. Simulated leather cover on plaid material bag, 1950s. (G.S.)

Value $65-110

2. Roy Rogers and Trigger illustrated images on simulated leather small bag with plastic shoulder straps. Acme Briefcase Company, 1950s. (G.S.)

Value $65-110

Bottom Row:

1. Dale Evans red 14″ x 9″ school bag with shoulder strap was available with handle also made of "Texon" material by United Leather Goods, late 1950s. (G.S.)

Value $150-300

2. Roy Rogers and Trigger tan "football grained" imitation leather school bag 10″ x 14″ with plastic handle and front zippered pencil case. Notice illustration of Roy on a pinto-colored Trigger. (G.S.)

Value $125-250

Roy Rogers red pencil box shown with Roy Rogers and Trigger pennant. #1 Roy Rogers pencil box slides open to reveal contents of box that includes ruler, crayons, pencils, pencil sharpener, and Pledge to Parents card. Roy and Trigger embossed in silver on bright red box, 1955-1956, Eagle Pencil Company, originally cost 29¢. (R.L.)

Value $90-150

Roy Rogers and Trigger 28" pennant, 1940s. (B.W.)

Value $75-125

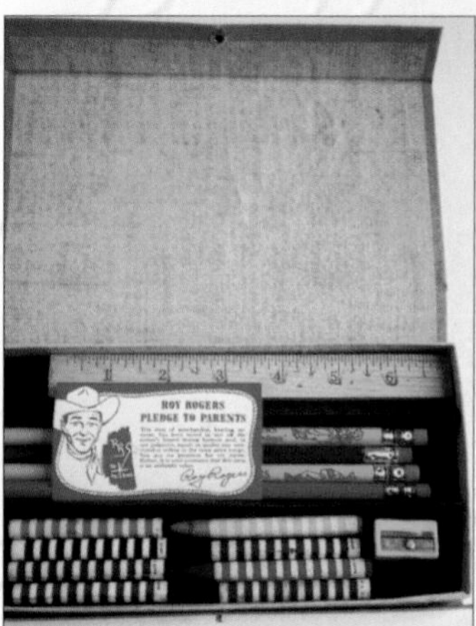

Roy Rogers Double R Bar Ranch pencil box, illustrated and embossed cover with Roy, Dale, Trigger, and Bullet shown in ranch scene. Roy Rogers brand logo on bottom corner. Eagle Pencil Company, 1955-1956, 8 1/2" x 4 3/4" inside of box contains ruler, pencils, pencil sharpener, crayons, and Pledge to Parents card. (M.M.)

Value $100-150

Roy Rogers Double R Bar Ranch pencil box. Top of tan box has black illustrated images of Roy and Trigger, Roy Rogers brand logo, Dale Evans, and Bullet. Eagle Pencil Company opened box has two tiers, 8 1/2" x 4 3/4" box contains pencils, ruler, and Pledge to Parents card. (D.T.)

Value $90-135

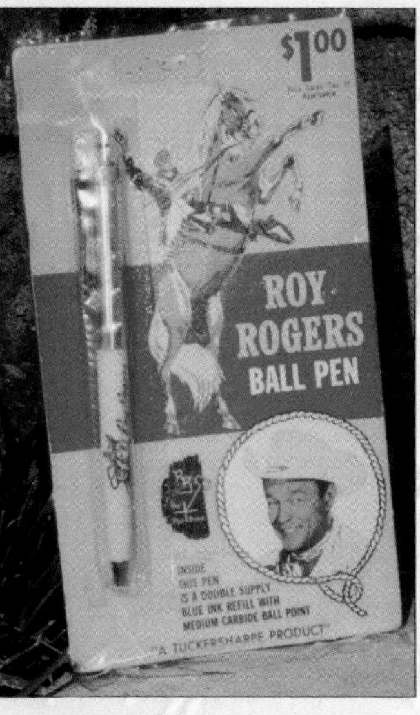

Roy Rogers ballpoint pens with different colored barrels, "A Tuckersharpe Product," early 1950s. (B.W. and M.M.)

Value $65-125

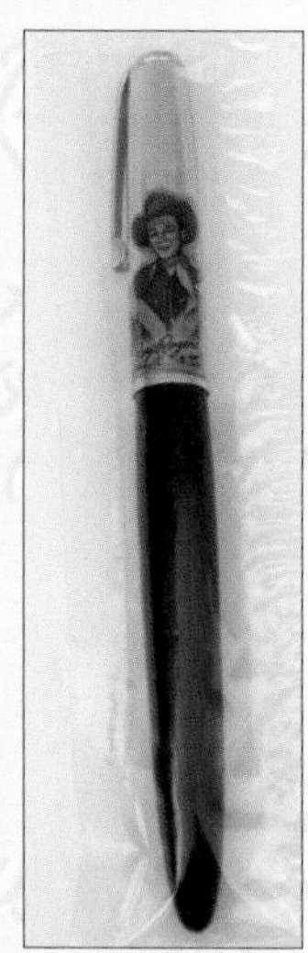

Roy Rogers pens made by Stratford Pen Corp. colored illustration of Roy on cap of pen. Roy's name on barrels, 1955-1956. (G.S.)

Value $65-125 on original card

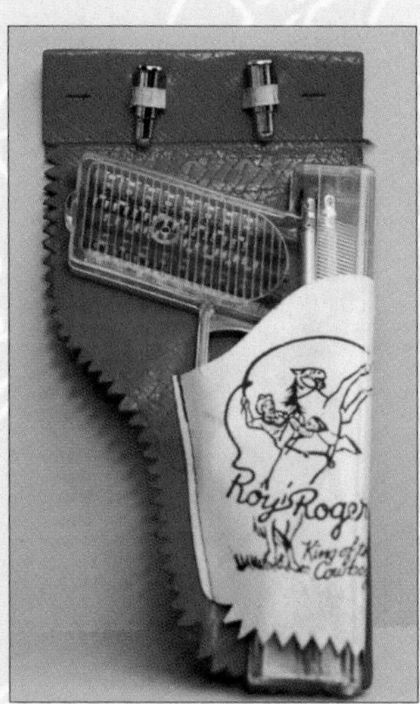

Roy Rogers "King of the Cowboys" plastic gun style pencil case with holster. Clear plastic gun contains crayons, pencils, and pencil sharpener. Red and white holster with silver colored plastic bullets is 10" long. (C.Q.)

Value $75-150

Roy Rogers "King of the Cowboys" clear plastic pencil case with brown and tan leatherette plastic holster. (M.M.)

Value $75-150

Roy Rogers vinyl pencil case with Roy on rearing Trigger, case came in colors of red, brown and green, case border has facsimile stitching. (C.Q.)

Value $50-75

Roy Rogers paper binders with textured simulated leather front and back covers, 10″ x 12″ illustration of Roy on rearing Trigger with his name in rope lettering of lasso. Came in brown, red, and green. (Roy Rogers Museum and D.T.)

Value $75-125

EDITION NUMBER TWO

My Weekly Reader®

Vol. XXIII | Week of April 19-23, 1954 | No. 28

News About Roy Rogers and Trigger

Roy Rogers went to England
this spring.
He went to give cowboy shows.

Dale Evans went with Roy.
So did Trigger.
Roy could not give a show
without good old Trigger.

Trigger knows 50 tricks.
Some people call him
the smartest horse in the world.

Roy Rogers leads Trigger from an airplane.

Roy says, "I have to be smart
to know Trigger's tricks.
Some children like Trigger more
than they do me.
They write letters to Trigger.
I am glad that they like my horse."

The children in England like
American cowboys.
They read stories about cowboys.
They see Roy Rogers and other
cowboys in the movies.

◄ Roy Rogers and Trigger

Roy Rogers and Trigger featured on cover of *My Weekly Reader* 1954, rare. (G.S.)

Value $25-50

Gold-plated metal statuette of Trigger, award was presented to school with the best safety record, 1950s, very rare. (G.S.)

Value $500-750

Roy Rogers and Trigger featured on cover of loose-leaf filler paper 8″ x 10″ package, 1950s. (M.M.)

Value $30-50

Gabby Hayes with facsimile signature featured on 8″ x 10″ tablet. (G.S.)

Value $30-50

Roy Rogers with Trigger's saddle featured on cover of 8″ x 10″ tablet. Lakeside Central and Southern Central Companies. (Roy Rogers Museum)

Value $30-50

Inside cover of Roy Rogers paper binder with package of notebook filler paper. (D.T.)

Package of notebook filler paper for binders, package label featured pictures of various movie stars. (D.T.)

Value $25-50

Collection of tablets featuring Roy Rogers, Dale Evans, Trigger and Buttermilk on front covers in 8″ x 10″ size made by Lakeside Central Company and Southern Central Company, 1953-1960. (G.S.)

Value $30-60 each

LUNCH BOXES AND THERMOSES

The American Thermos Bottle Company was the leading manufacturer of Roy Rogers and Dale Evans lunch boxes and thermoses. They produced the first lithographed Roy and Dale lunch kits in 1953, and sold two and a half million in their first year of manufacture. The first completely lithographed steel lunch box and thermos artwork was done by Ed Wexler. He had a great reputation for illustrating the likenesses of Roy and Dale. The fully illustrated box became a standard for all lunch boxes made between 1953 and 1960.

During this time period the American Thermos Bottle Company produced lunch boxes with 12 different Roy and Dale designs and a variety of thermoses and red plastic thermos cups. The American Thermos Company changed its name to King Seeley Thermos in 1960 and produced the only Roy Rogers saddlebag-type vinyl lunch box.

Roy Rogers and Dale Evans lunch box came in wood grain, red and blue band versions, and with leather or red plastic handles. Front of box is illustrated with Roy riding Trigger and Dale with Bullet in background scene. Back of box has Roy Rogers encircled with rope design 8 1/2" x 6 1/2" x 4" by American Thermos Co., 1950s. (C.Q.)

Value $175-325

Roy Rogers vinyl lunch box made in 1960 by King Seeley Thermos Co. 9" x 7" x 4" was produced in white and tan and came with yellow sky thermos with Roy on rearing Trigger. (C.Q. and G.S.)

Value $200-450

Roy Rogers and Dale Evans domed Chow Wagon lunch box, made 1958-1961 by the American Thermos Company and came with yellow sky thermos, Dale illustrated on end of box and Roy on the side. (C.Q.)

Value $250-450

Roy Rogers on rearing Trigger thermos that came with eight-scene lunch box, blue-sky background, by American Thermos Co. (C.Q.)

Value $75-150

Roy Rogers and Dale Evans lunch box came in red and green band versions with metal or red plastic handles. Front of box shows eight illustrated scenes of Roy, Dale, Trigger, and Bullet. Other side of box has Roy on rearing Trigger surrounded by cattle, size 8 1/2" x 6 1/2" x 4" by American Thermos Co. 1950s. (C.Q.)

Value $200-350

Roy Rogers and Dale Evans lunch box 8 1/2" x 6 1/2" x 4" with illustrations of Roy, Dale, and Bullet on front and Roy roping calf on back, blue border on both sides. (C.Q.)

Value $200-350

Roy Rogers on rearing Trigger thermos with yellow sky background. (C.Q.)

Value $75-150

CHAPTER 20

STORE AND COUNTER DISPLAYS

The birth of a merchandizing empire for Roy Rogers occurred in 1940 when Republic Pictures put a clause in his movie contract that allowed him to pursue commercial avenues. The timing was perfect for Roy as the popularity for cowboy movies among the public was increasing. Products endorsed by Hopalong Cassidy were already selling well in retail stores.

Manufacturing companies jumped at the idea of putting Roy's character name and image on their products, and by 1948, with the aid of a huge national advertising campaign, Roy Rogers merchandise was flying off the shelves in large retail stores. The Sears, Roebuck & Company catalog and stores carried every type of Roy Rogers item from furniture to toothbrushes. Macy's featured Roy Rogers Corrals in various departments of their stores during the 1950s, and sales for both companies soared during this time period. By 1960, Roy Rogers movies, comics, personal appearances, and the merchandizing empire came to a halt.

Photograph of Roy Rogers Ranch Club furniture store display, 1950s. (Roy Rogers Museum)

STORE DISPLAYS

Original candid photograph of Roy Rogers conferring with men about products bearing his name. (Roy Rogers Museum)

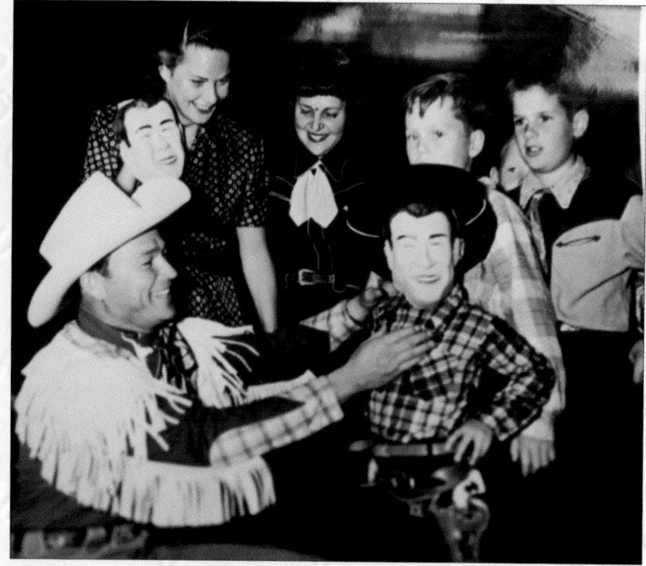

Original candid photograph of Roy Rogers and child wearing a Roy Rogers mask and western outfit with other children looking on. (Roy Rogers Museum)

Original candid photograph of Roy Rogers looking over a group of Roy Rogers endorsed items. (Roy Rogers Museum)

Photograph of Roy Rogers cowboy boots store display, 1950s. (Roy Rogers Museum)

Photograph of Roy Rogers gun and holsters store display. (Roy Rogers Museum)

Original candid photographs of Macy's Roy Rogers Western Corrals featuring endorsed merchandise, 1950s. (Roy Rogers Museum)

Original candid photograph of Roy Rogers with children's mechanical Trigger Ride, 1950s. (Roy Rogers Museum)

Photograph of Roy Rogers records, retail store window display, 1950s. (Roy Rogers Museum).

Roy Rogers Rodeo Rides coin-operated rides by Ri-Deo-Kiddielane Corp. Retail merchants catalog advertisements, 1950s. (Roy Rogers Museum)

COUNTER DISPLAYS

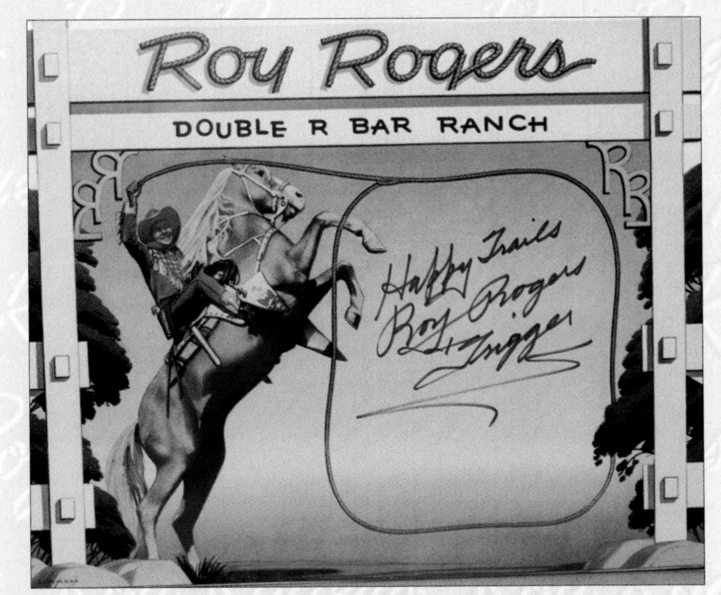

Roy Rogers Double R Bar Ranch counter display 20" x 16" used to display various Roy Rogers items, 1950s. (C.Q.)

Value $350-550

Roy Rogers "Win a Roy Rogers Pony!" 1" x 7 1/2" store counter display with entry form, Krinkles Cereal. (M.M.)

Value $85-150

Roy Rogers Straight-Shooter gun puzzle key chain counter display with 24 plastic puzzle guns, extremely rare. The Plas-Trix Co. (L.C.)

Value $2,200-3,000

Roy Rogers Riders harmonicas store counter display, Harmonic Reed Corp., 1950s. (G.S.)

Value $1,200-1,800

Roy Rogers and Dale Evans free-standing Kool-Aid counter display, extremely rare. (G.S.)

Value Unknown

Roy Rogers and Trigger Jack Knife counter display, 1950s. (B.W.)

Value $200-350 without jack knife

Roy Rogers Ball point Pen store counter display with pens on display cards, "A Tuckersharpe Product," 1950s. (G.S.)

Value $250-500

Roy Rogers trick lasso counter display, electric motor behind cardboard photograph of Roy moves lasso, extremely rare. (G.S.)

Value $800-1,200

Original photograph of Roy Rogers and Trigger store counter displays, 1950s. (Roy Rogers Museum)

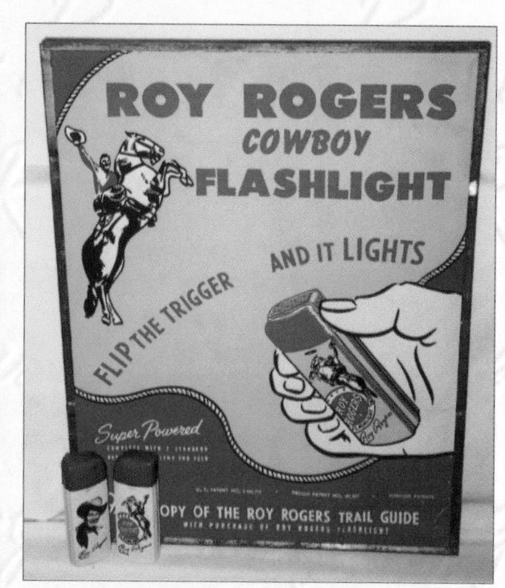

Roy Rogers Cowboy Flashlight 10" x 12" counter display sign, Bantam-Light, Inc., 1950s. (G.S.)

Value $150-250

Roy Rogers and Trigger counter display, full color illustration of Roy on rearing Trigger, used to display smaller items like watches. Made of fiberboard, extremely rare. (B.W. and C.Q.)

Value $600-1,200

Roy Rogers Tribute counter display for CDs, 24" high, made recently. (C.Q.)

Value $45-90

Roy Rogers Cowboy Flashlight 9 3/4" x 10" counter displays by Bantam Light, Inc. Trail guide was free with purchase of pocket-size flashlight. (C.Q. and B.W.)

Value $450-750 complete with flashlights

CHAPTER 21

TOYS

Roy Rogers' huge popularity in the 1940s and 1950s was a major force that helped retail stores sell thousands of toys and other items bearing his name and image. Toy manufacturers helped pay for large one-page full-color Roy Rogers toy advertisements produced for Macy's and Sears, Roebuck & Company retail stores. These advertisements were placed in major national magazines and were another positive force in selling toys. Most of these ads were run right before Christmas, as toys always topped children's Christmas wish list.

Little buckaroos felt closer to their hero by playing with Roy Rogers toys. They could strap on a pair of Roy's toy six guns and bring the bad guys to justice just like Roy did in the movies and on his weekly television show. Almost every type of toy was manufactured with the "King of the Cowboys" name on it for children to play with in the wild west world of Roy Rogers and his co-stars.

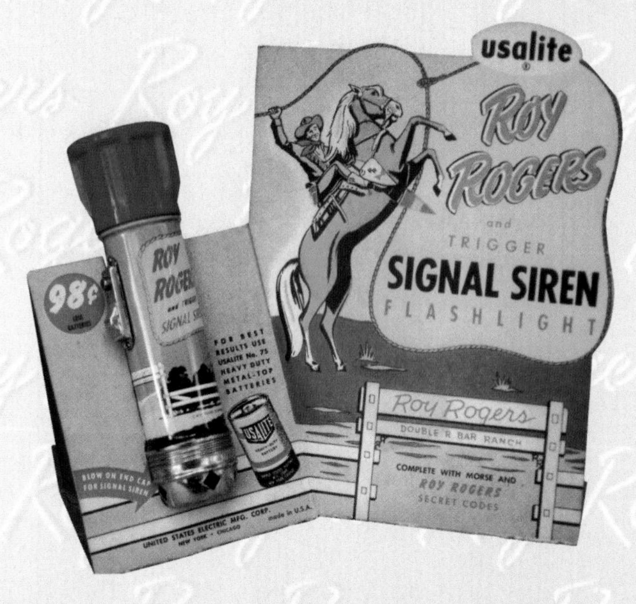

Roy Rogers and Trigger signal siren flashlight with original box, secret code booklet and pledge to parents card. 3-way switch, siren in cap. Illustrated Roy on rearing Trigger in full color on 6" long metal case. Made by Usalite, 1954. (C.Q.)

Value $175-250

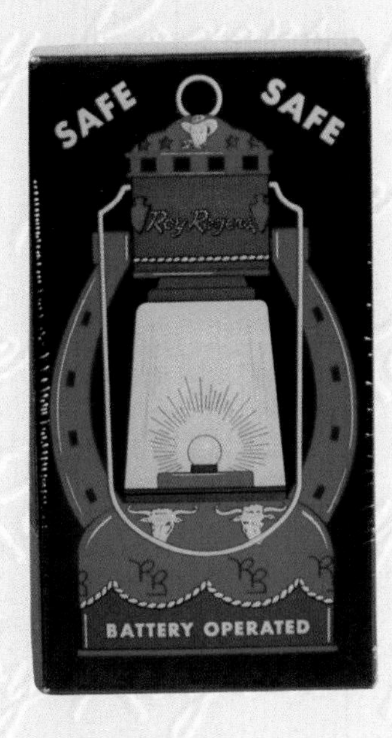

Roy Rogers ranch lantern with original box by Ohio Art Co. Battery powered, 8 1/2" tall. (C.Q.)

Value $125-225

Gabby Hayes jailer keys by John Henry Products on original display card. (M.M.)

Value $65-125

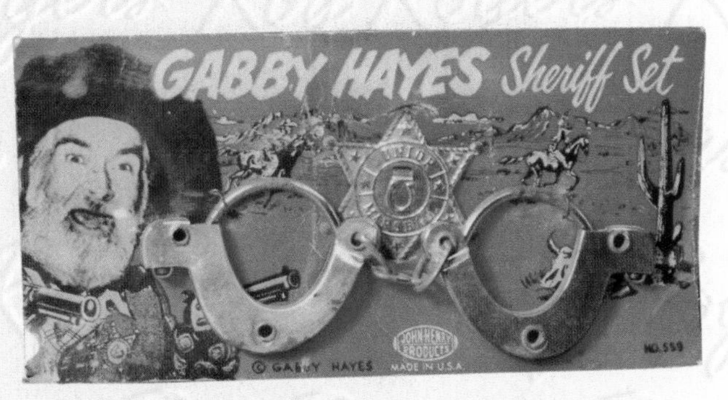

Gabby Hayes sheriff set by John Henry Products on original display 4 1/2" x 9 1/2" card. (C.Q.)

Value $65-125

Roy Rogers cowboy flashlight with original box, 2 1/2" long. Made by Bantam-Lite Inc., 1950. (D.T.)

Value $45-90

Gabby Hayes miniature old-time automobiles with order form and shipping box, rare. (M.M.)

Value $90-150

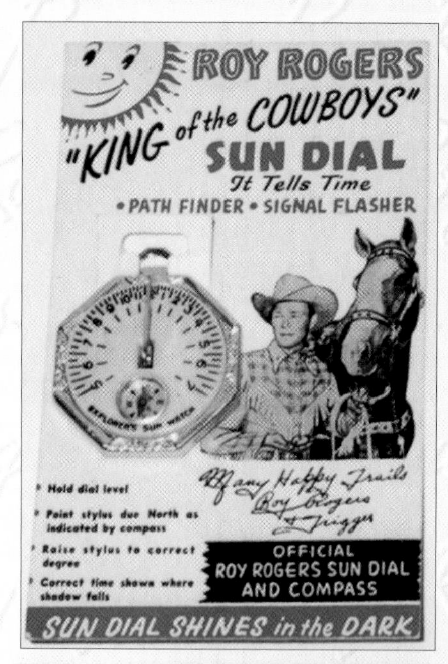

Roy Rogers "King of the Cowboys" sun dial on original display cards, rare. (G.S.)

Value $150-250

Roy Rogers and Trigger marbles, 3/4" diameter. (C.Q.)

Value $5-10 each

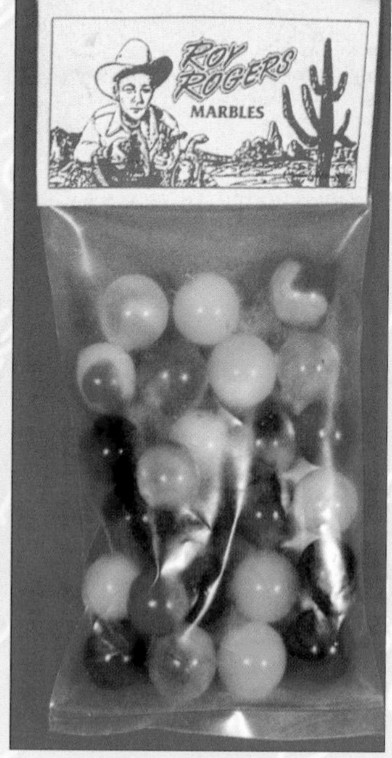

Roy Rogers marbles in original display 3" x 6" bag, 1950s. (C.Q.)

Value $30-50

Roy Rogers "The King of the Cowboys" 1940s marbles in original bag. (C.Q.)

Value $25-40

Roy Rogers trick lasso with display package and photograph of Roy. Glows in dark. Made by Knox-Reese Co. Sold at Rodeos and offered as prize in National Lasso Contest, 1952. (C.Q.)

Value $125-175

Roy Rogers rodeo lariat with original display tag. Came in 2 sizes, 23" diameter and 28" diameter. 3/4" thick manila hemp rope. Made by Lareo Company, Inc., 1959-1960. (C.Q.)

Value $175-250

Roy Rogers branding iron set in original box with ink pad. Made by Knox-Reese Co., 1950s. (C.Q.)

Value $125-175

Roy Rogers hauler, van trailer and Nellybelle Jeep with original box by Louis Marx and Company. Made of steel, doors open on trailer to form ramp. Rare with box. 15 1/2" long, 1955-1956. (G.S.)

Value $450-900

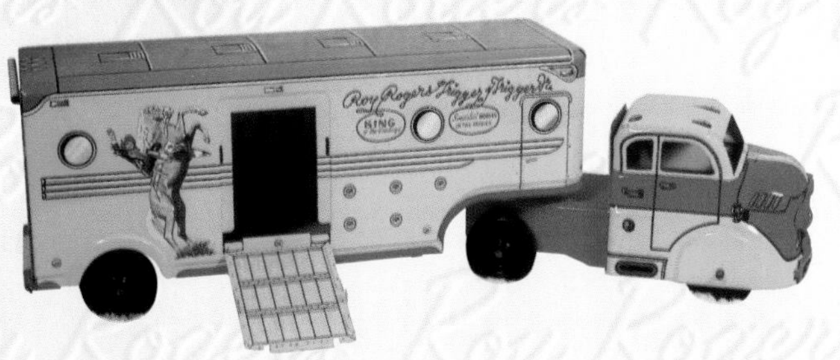

Roy Rogers hauler and trailer with original box, Louis Marx and Company. Came with Roy, Dale, Pat Brady and Bullet plastic figures plus two plastic horses with removable saddles and bridles. Rare with box, 1955-1956. (G.S.)

Value $450-900

Trigger wood pull toy by N. N. Hill Brass Co., 16" long with metal wagon wheels, 1950s. (C.Q.)

Value $225-425

Roy Rogers hauler and van trailer with original box. Operates with battery powered remote control, 13" long. Made by Line Mar Co., very rare, 1957. (G.S.)

Value $450-900

Roy Rogers and Trigger pull toy with box. Made by N.N. Hill Brass Co. 1950s. (C.Q.)

Value $200-355

Nellybelle Jeep. 5" x 5" x 11". Hood raises up, side panels are removable, detailed engine. Came with plastic figures of Dale Evans, Pat Brady, and Bullet. Gray pressed steel. Made by Louis Marx Co., mid-1950s. (C.Q.)

Value $175-300 Add $200 for figures and original box

Nellybelle Jeep and trailer with Roy Rogers, Pat Brady, and Trigger plastic figures. 14 1/2" long, 1954. (C.Q.)

Value $250-350

Roy Rogers western telephone with original box, 9" high. Brown plastic with black plastic ear and mouthpieces and brass plated metal ringers. Made by Ideal Company, 1950s. (C.Q.)

Value $175-250

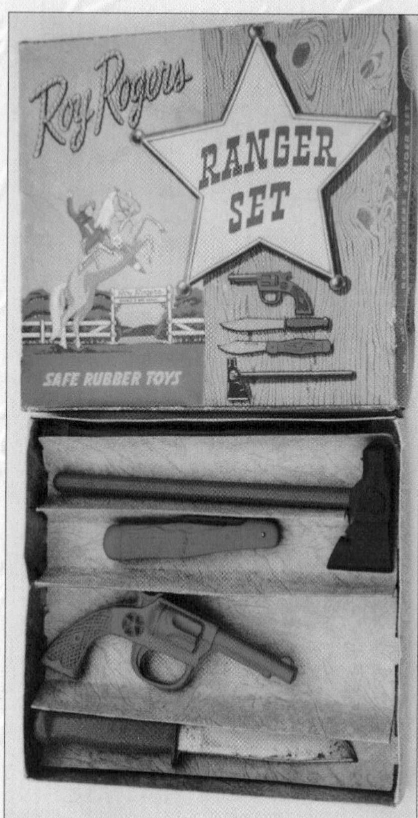

Roy Rogers ranger set made by Auburn Rubber Co. with rubber knives, gun, and hatchet. 1950s. Rare. (G.S.)

Value $250-350

Roy Rogers and Dale Evans hand puppets. Rubber flesh colored heads, of Roy and Dale with cloth hands and shirts, 8" long, 1950s. (G.S.)

Value $90-175 each

Dale Evans and Gabby Hayes hand puppets. Rubber heads with cloth hands and shirts, 1950s. (C.Q.)

Value $90-175 each

Roy Rogers hand puppet with original box. Made by Zany Toys, 1950s. (C.Q.)

Value $90-175

Roy Rogers and Trigger football. White molded rubber with black lace pebble-grained, regulation size. J.A. Dubow Sporting Goods Co., 1955-1956. Rare. (C.Q.)

Value $150-250

Roy Rogers and Trigger football and basketball. Brown pebble-grained rubber with black illustration of Roy on rearing Trigger. J.A. Dubow Sporting Goods Co., 1955-1956. (M.M.)

Value $150-250 each

Gabby Hayes fishing outfit with original illustrated tube container. Reel with 2-piece rod. Included Gabby Hayes adventure comic, 1950s. Rare. (M.M. and G.S.)

Value $350-600

Roy Rogers archery set on original display board. Rubber tipped cedar wood arrows. Came with 39" or 38" hickory bow and instruction booklet. Made by Ben Pearson, Inc., 1953-1956. Rare. (B.W.)

Value $150-300

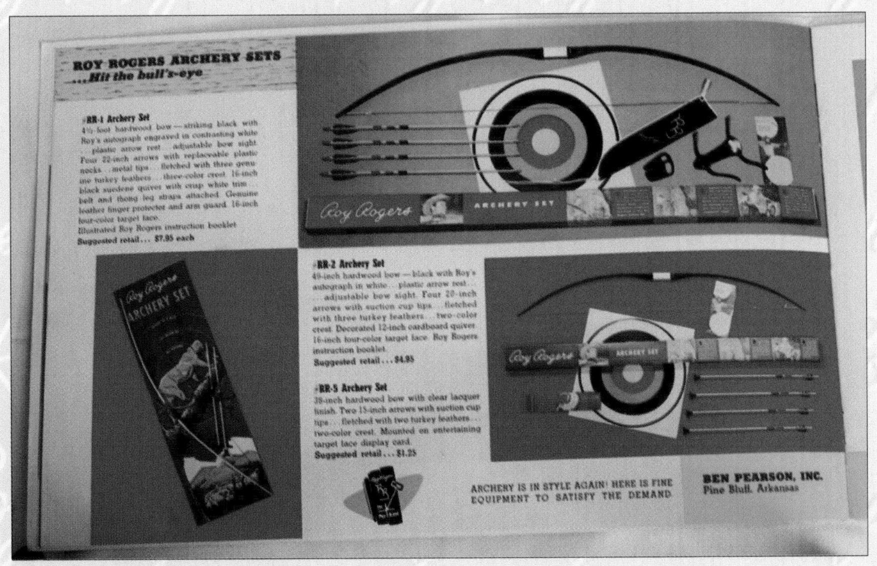

Retail Merchants catalog advertisement for Roy Rogers archery sets made by Ben Pearson, Inc. 1950s. (Roy Rogers Museum)

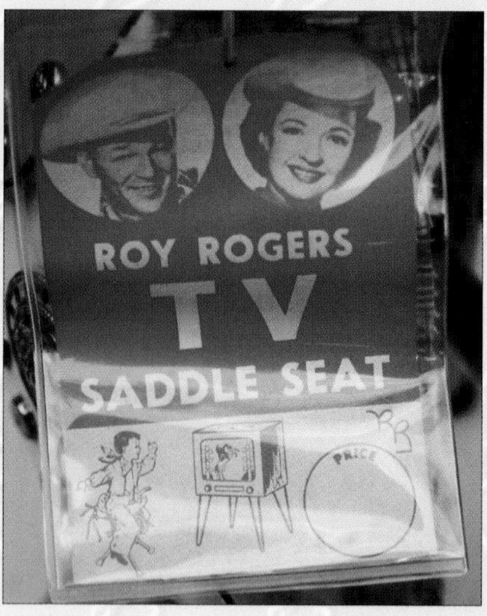

Original tag on Roy Rogers and Trigger TV Saddle Seat. (C.Q.)

Roy Rogers and Trigger TV Saddle Seat. Leather saddle seat with 3-leg wood frame. Made by Haskell Benson Company, 1950s. (C.Q.)

Value $350-500

Trigger wood and metal rocking horse. (C.Q.)

Value $300-500

Roy Rogers Trigger toddlers riding horse by Suzy Goose, 24″ high. Steel frame with plastic head, seat and wheels, 1950s. (C.Q.)

Value $300-450

Roy Rogers Trigger Play Horse 1958. 20″ high with 18 1/2″ wheel base. Plush toy with plastic seat and metal wheels. The Stern Toy Company. (R.L.)

Value $250-350

Original photograph of Trigger and Bullet plush toys made by the Stern Toy Company. (Roy Rogers Museum)

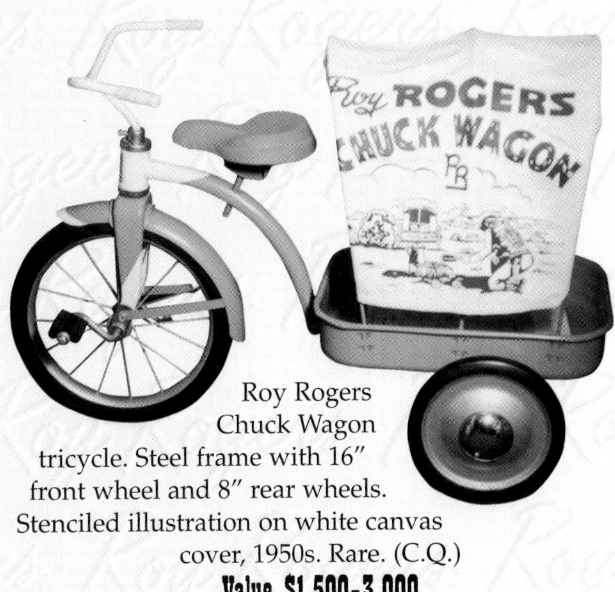

Roy Rogers Chuck Wagon tricycle. Steel frame with 16″ front wheel and 8″ rear wheels. Stenciled illustration on white canvas cover, 1950s. Rare. (C.Q.)

Value $1,500-3,000

Roy Rogers Chuck Wagon, 17″ x 40″ with steel frame, pull handle and wheels. Wood bed with canvas cover. Made by Steger Company, 1950s. Rare. (B.W.)

Value $850-1,500

Roy Rogers Toy Wagon. Wood with illustrated canvas cover, 16″ x 27″. (C.Q.)

Value $300-500

Nellybelle Jeep pedal car, 29 1/2″ long. Steel body, frame, and wheels with ball bearing rear axle. Illustration of Roy on rearing Trigger and Nellybelle name on doors. Hamilton Steel products, Inc. (Roy Rogers Museum)

Value $3,500-5,500

Original candid photograph of Pat Brady and Nellybelle. (R.L.)

Roy Rogers buckboard wagon. Shown without side boards. 17″ x 40″ with 10″ wheels. Steel frame, wheels and handle with wood bed, 1950s. (B.W.)

Value $650-900

Roy Rogers bunkhouse tent. Shown not set up. Extremely rare. (G.S.)

Value $750-1,200

Original candid photograph of children in Roy Rogers and Trigger tent. 1950s. (Roy Rogers Museum)

Roy Rogers Redwood yellow teepee tent. Extremely rare. (C.Q.)

Value $750-1,200

Roy Rogers teepee tent scale model salesman sample. Extremely rare. (L.C.)

Value $700-950

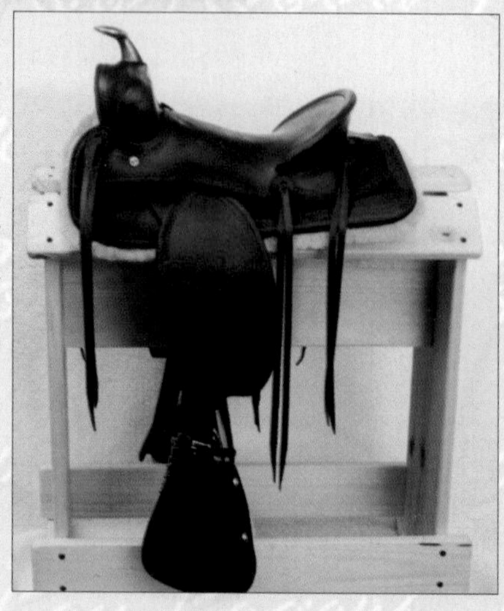

Roy Rogers children's saddle. Made by Bona-Allen Co. in black leather with Roy's name and horse with rider embossed on fenders. Rolled rope border, 12″ seat, 1952. Rare. (C.Q.)

Value $900-1,800

HARTLAND FIGURES

Original candid photograph of prototype Roy Rogers, Dale Evans, and Bullet plastic figures made by Hartland Plastics, Inc. (Roy Rogers Museum)

Roy Rogers and Trigger Hartland Figures with original box. Horse and rider series by Hartland Plastics, Inc. (C.Q.)

Value $200-300

Roy Rogers and Dale Evans illustrated on top lid of Hartland Plastic, Inc. 1950s box. (C.Q.)

Value $150-200 box only

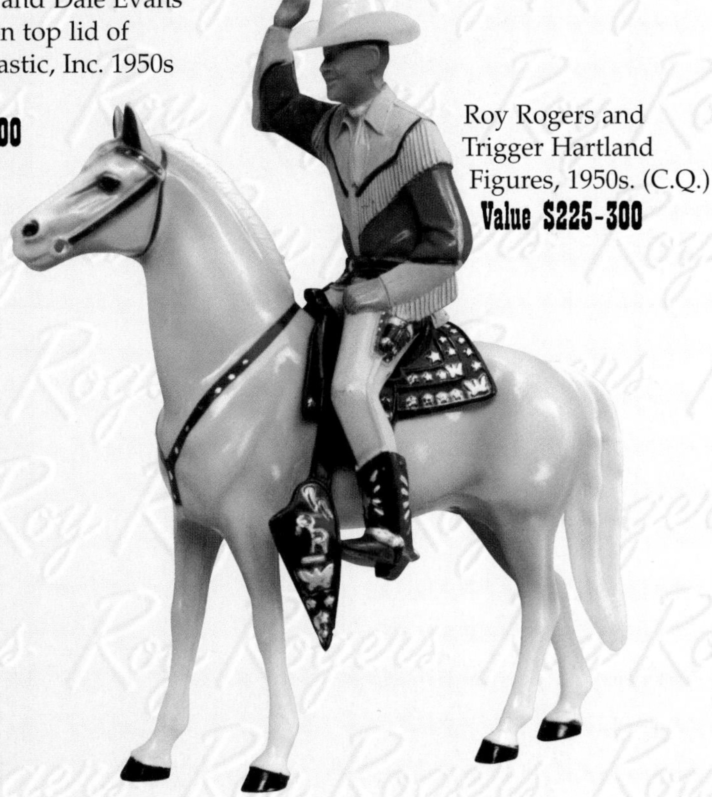

Roy Rogers and Trigger Hartland Figures, 1950s. (C.Q.)

Value $225-300

Roy Rogers and Trigger Hartland figures with original hang tag and box, 1950s. (M.M.)

Value $250-350

Dale Evans and Buttermilk hang tag from Hartland plastic figure. (C.Q.)

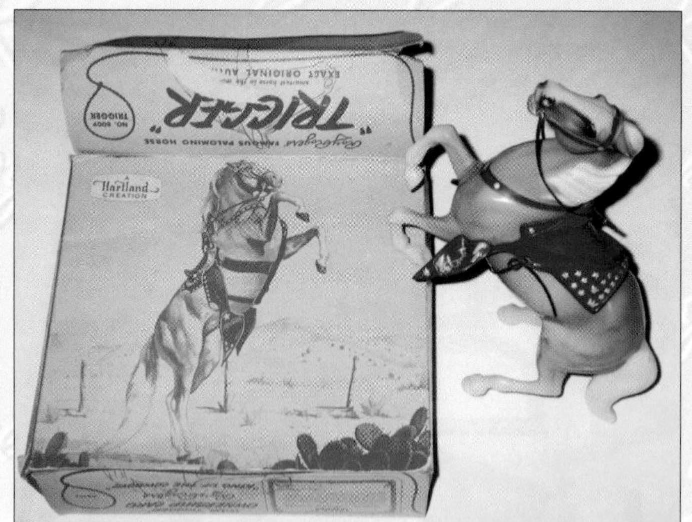

Hartland Trigger with rare original box by Hartland Plastics, Inc., 1950s. (G.S.)

Value $300-450

Hartland Bullet with rare original box, 1950s. (C.Q.)

Value $300-450

Roy Rogers and Trigger Hartland Figures, 1950s. (C.Q.)

Value $225-300

PLAY SETS

Retail merchants catalog advertisements for Roy Rogers ranch and town sets made by Louis Marx and Company, Inc., 1950s. (Roy Rogers Museum)

Roy Rogers Double-R-Bar ranch set by Louis Marx and Co., set #3980. Lithographed tin bunkhouses with plastic fences and figures, 1950s. (Roy Rogers Museum)

Value $250-350, add $100-150 for original box

Original Roy Rogers rodeo ranch set boxes for two different sets, #3990 and #3995. Top box measures 4″ x 9″ x 22″ and sets include ranch house, white plastic rail fences, 2 1/2″ tall plastic figures of Roy and nine other cowboys. Also includes livestock figures and other items. Complete sets. (G.S.)

Value $350-550

Roy Rogers Western Town Set with original box, box measures 6″ x 10″ x 34″. Lithographed tin Mineral City buildings. Series 5000, set #4258. Set includes plastic figures and other items. Marx Co., 1950s. (G.S.)

Value $450-700

Roy Rogers cabin construction set made by Lincoln Logs Co. Set includes cardboard figures, 1950s. Rare. (G.S.)

Value $350-500

Roy Rogers with Trigger, cowboys, horses, and accessories, 27-piece set by Louis Marx and Co., 1956. (C.Q.)

Value $100-175

Roy Rogers stage coach wagon train made by Louis Marx Co. with rare box. Red version shown, it also came in blue, 1950s. (M.M.)

Value $450-600

Roy Rogers Fix-It chuck wagon and jeep by Ideal Toy Co. with original box, 4 1/2″ x 7″ x 24″. Plastic wagon, jeep, and figures,1950s. (C.Q.)

Value $250-400 complete set

Roy Rogers Fix-It Chuck Wagon and Jeep by Ideal Toy Corp. Set made with different color chuck wagons.

Roy Rogers Fix-It stagecoach made by Ideal Toy Co. 5″ x 6″ x 15″. Plastic stagecoach and figures, 1955. (R.L.)

Value $150-200, add $125 for box

BANKS

Roy Rogers metal boot bank made by Almar Metal Arts Co., 3 1/2″ x 4 1/2″ with original box. Painted to look like real boot, 1950s. (G.S.)

Value $125-225

Roy Rogers and Trigger savings bank made by Ohio Art Co. Lithographed tin bank is 8″ x 6″ and mounts on the wall, 1950s. (C.Q.)

Value $150-250, add $100 for original box and key

Roy Rogers metal boot bank by Almar Metal Arts Co. Copper plated with original box. Roy on rearing Trigger on sides of boot. 3 1/2″ x 4 1/2″. (G.S. and C.Q.)

Value $125-225

Roy Rogers on rearing Trigger ceramic bank. 7 1/2″ tall. Very rare. (C.Q.)

Value $125-200

Roy Rogers ceramic bank, 6″ tall, 1950s. (C.Q.)

Value $60-125

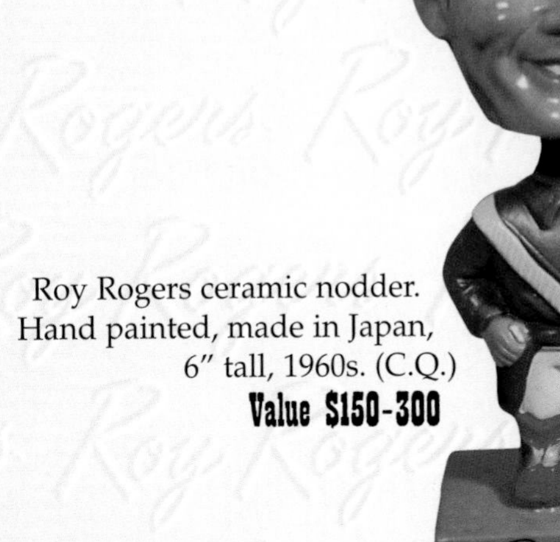

Roy Rogers ceramic nodder. Hand painted, made in Japan, 6″ tall, 1960s. (C.Q.)

Value $150-300

OPTICAL

Collection of Roy Rogers binoculars and cameras mint in the original boxes. Very rare boxes. (G.S.)

Roy Rogers 3-power Binoculars with original boxes in 3 different colors, 4 1/2″ x 4″. Made by Herbert George Co., 1950s. (C.Q. and G.S.)

Value $125-175, add $100 for box

Roy Rogers flash camera made by Herbert George Co., 1950s. (C.Q.)

Value $100-150

Original photograph of Roy Rogers binoculars, camera, and telescope made by Herbert George Co. Notice that no decals are on binocular or telescope. 1950s. (Roy Rogers Museum)

Roy Rogers 3-power binocular with original box. Made by Herbert George Co. Pebble finish, illustrated decals of Roy and Roy on rearing Trigger, 5" x 4 1/2". Extremely rare. (G.S.)

Value $250-400

Roy Rogers camera and binocular set made by Herbert George Co. with original rare box, #620 snap-shot camera, plastic and metal, 1950s. (G.S.)

Value $250-400

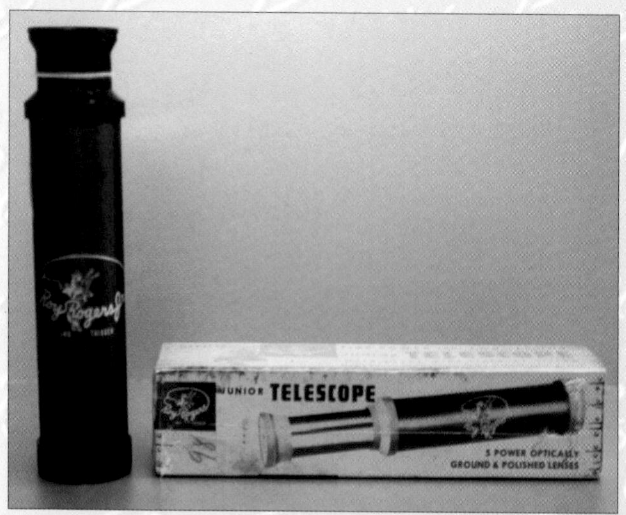

Roy Rogers 5-power telescope with original box made by Herbert George Co. Plastic with metal trim, 1950s. Rare. (C.Q.)

Value $175-300

Roy Rogers photographic kit with original rare 2-piece box, insert, and camera bag. Kit included. 12 pieces. Made by Herbert George Co., 1950s. (G.S.)

Value $300-500

INSTRUCTIONS FOR USE AND CARE

VIEW FINDER

TAKING LENS

LOADING THE CAMERA

Place this roll of film in the lower portion of the focal box opposite the winding knob. This is easily done by fastening the spool securely into the two pivots. Be sure that the black side of the film is downward and the roll is placed over the retainer spring. You will see an empty spool on the opposite side held securely by the winding knob. To be absolutely sure, turn the winding knob several times to see that the empty spool turns with the knob.

Thread the pointed end of the paper into the large slot of the empty spool. Make sure the paper is centered on the empty spool in order that it will wind evenly. Now give the winding knob about one full turn—just enough to bind the paper and assure a firm grip on the spool.

WINDING THE FILM

You are now ready to replace the front and the back of the camera in its proper place. You will note a slot in the upper left hand of the back—be sure that the knob slips into place into this slot. Replace both clips over retaining lugs, and be sure that camera is locked securely.

At the back of camera you will note a red window. Turn winding knob until hand appears then turn until Number 1 appears. You are now ready to take your first picture with your Roy Rogers Camera.

TO TAKE A PICTURE

Hold camera securely in your left hand and bring telescopic view finder to your eye. Frame the subject you wish to photograph in the view finder. The subject should not be closer than eight feet from the camera. A single downward or upward stroke of the shutter release operates the shutter and snaps the picture. After the first picture has been taken turn the winding knob until number 2 appears in the red window. Repeat this operation until twelve pictures have been snapped.

REMOVING THE FILM

IMPORTANT THINGS TO REMEMBER

Roy Rogers camera instruction booklet. Came with Herbert George Cameras, 1950s. (C.Q.)

Value $15-20

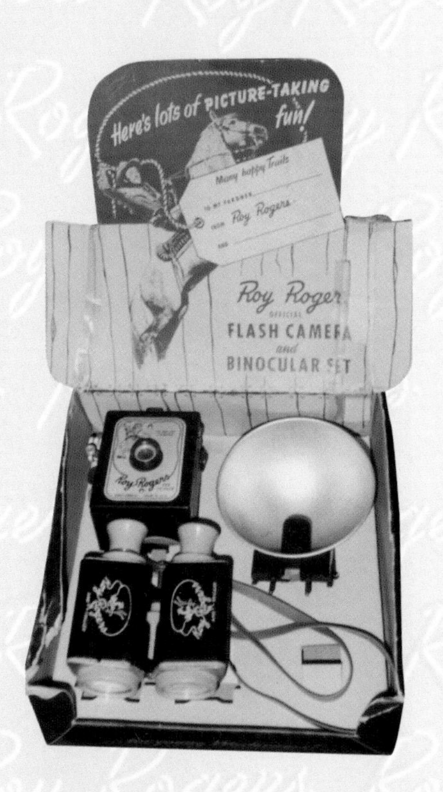

Roy Rogers flash camera and binocular set with rare original box. Made by Herbert George Co., 1950s. (G.S.)

Value $300-500

Roy Rogers televiewers with premium order form. Plastic brown and gray cases. (M.M.)

Value $45-125 each

Original 8" x 10" black and white photograph of Roy Rogers with yo-yo, 1950s. (Roy Rogers Museum)

Original photos of prototype Roy Rogers and Dale Evans yo-yos made by Roundup King. (Roy Rogers Museum)

Original 8″ x 10″ black and white photograph of Roy Rogers and children playing with yo-yos. (Roy Rogers Museum)

Roy Rogers yo-yo store display box. Came with a dozen yo-yos made by Roundup King, 1950s. (C.Q.)

Value $150-250

Roy Rogers Rodeo game, 4 games in 1, made by the Rogden Co., 9 1/2″ x 19″, 1950s. (C.Q.)

Value $125-250

Roy Rogers and Trigger yo-yo with original wrapper. Plastic yo-yo came in many colors. Made by Roundup King. (C.Q.)

Value $15-30

Roy Rogers and Trigger Pitch-Em Cowboy game, 2 1/2″ x 5″, 1940s. (C.Q.)

Value $25-40

Roy Rogers and Trigger wood burning set. Made by Rapaport Brothers, Inc., early 1950s. (C.Q.)

Value $200-300

Gabby Hayes target with original box and pistol set. Plastic pistol shoots suction-cup darts on metal dartboard. Extremely rare. (G.S.)

Value $300-450

Roy Rogers darts dartboard. Made by Dartboard Equipment Co., early 1950s. (B.W.)

Value $150-250

Roy Rogers and Trigger horseshoe pitching set with box. Made by Knox-Reese Manufacturing Co. in early 1950s. Box measures 10 1/2" x 8 1/2". (C.Q.)

Value $150-250

Roy Rogers horseshoe set by Ohio Art Co. Set #531 contains two red and two black vinyl shoes, two lithographed tin bases, and two metal screw-on pegs, 1950s. (C.Q.)

Value $150-250

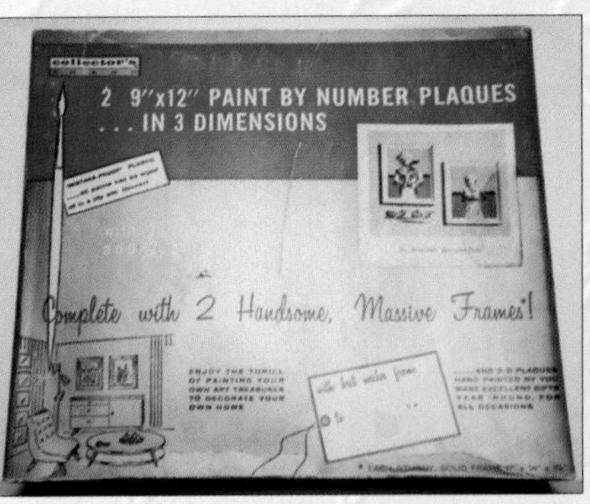

Roy Rogers and Dale Evans Paint by Numbers Plaques. Probably unauthorized as Roy and Dale's names are not on the set. Contains two 9" x 12" 3-dimenssional plaques, paint brush, and paints, 1950s. (M.M.)

Value $75-150

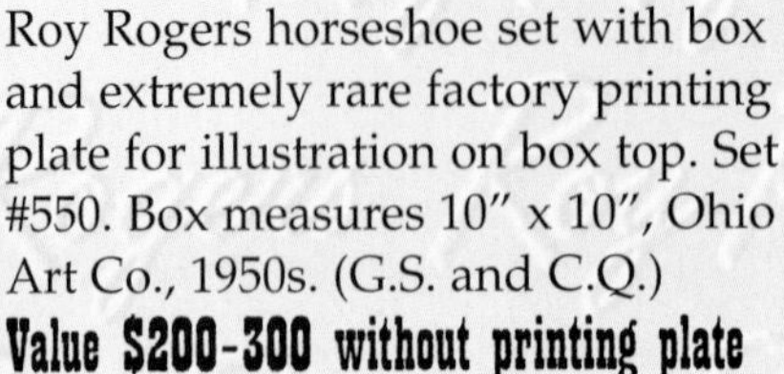

Roy Rogers horseshoe set with box and extremely rare factory printing plate for illustration on box top. Set #550. Box measures 10" x 10", Ohio Art Co., 1950s. (G.S. and C.Q.)

Value $200-300 without printing plate

Roy Rogers paint set. 13 3/4" x 9 1/4". Briefcase type box with metal handle with four paint-by-numbers pictures. Made by Standard ToyKraft Products, 1950s. (C.Q.)

Value $150-250

Roy Rogers and Dale Evans paint by number plaque set. Three-dimensional plaques with nine non-toxic paints, thinner, and brush. Four different pictures were produced, three shown are Roy on rearing Trigger, Roy feeding Trigger, and Dale riding Buttermilk. Not shown is picture of Roy and Bullet. Made by Collectors Craft Co. These sets were offered as prizes in Bakers' Instant Chocolate contest, 1957. Very Rare. (M.M.)

Value $125-200 each

Roy Rogers oil painting set, General Foods premium. Photo of Roy, Dale, and Trigger on box. Probably made by Standard ToyKraft Products. Three different sets made, 1953–1955. (D.T.)

Value $150-250

Roy Rogers oil painting set. Made by Standard ToyKraft Products, Inc., 1950s. (C.Q.)

Value $125-250

Roy Rogers modeling clay set. Made by Standard ToyKraft Products, Inc. 1950s. (C.Q.)

Value $125-250

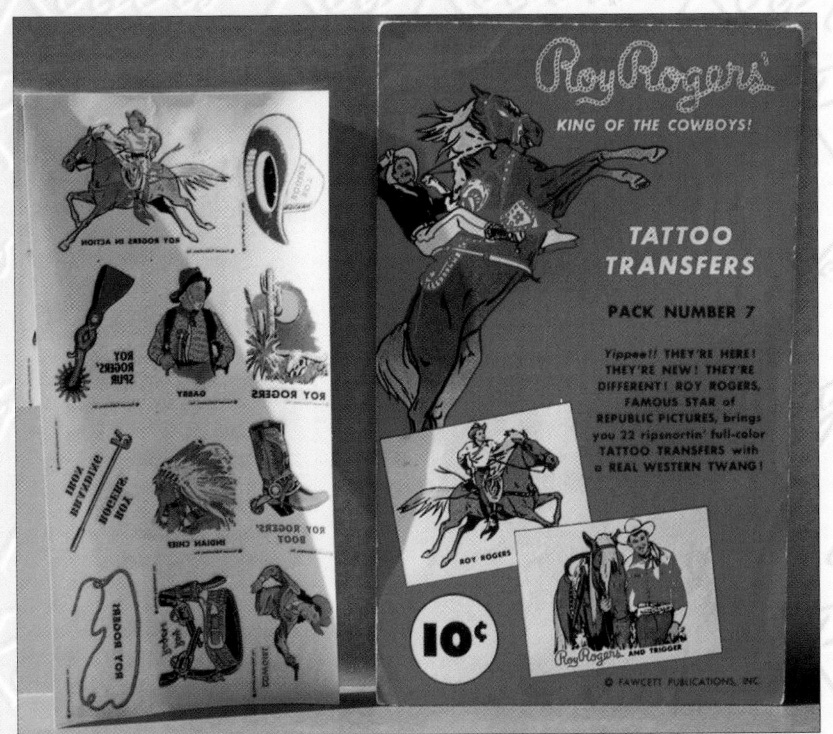

Roy Rogers tattoo transfers, 1950s. (C.Q.)

Value $45-90

PUZZLES

Whitman Publishing company produced a series of Roy Rogers frame tray inlay picture puzzles starting in the late 1940s and ending in 1958. Actual color photographs of Roy Rogers, Dale Evans, Trigger, Buttermilk, and Bullet were used to create these various sized puzzles. Many of these photographs were also used on the covers of Roy Rogers comic books.

Roy Rogers frame tray inlay picture puzzle with box, 11 1/2" x 14 1/2". Roy sitting with Trigger behind him, 1950s. (G.S.)
Value $50-85

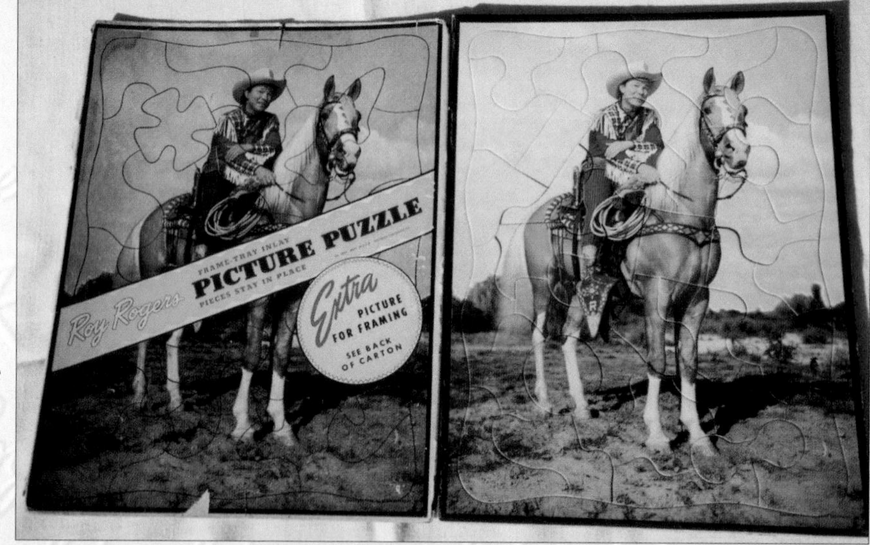

Roy Rogers frame tray inlay picture puzzle with box, 11 1/2" x 14 1/2". Roy sitting astride Trigger, 1950s. (G.S.)
Value $50-85

Roy Rogers frame tray inlay picture puzzle with box, 11 1/2" x 14 1/2". Roy leaning on saddle, 1950s. (G.S.)
Value $50-85

Roy Rogers frame tray inlay 9 1/2″ x 11″ picture puzzles, 1950s. (G.S.)
#1. Roy Rogers holding puppies.
#2. Roy Rogers and Bullet.
Value $35-50 each

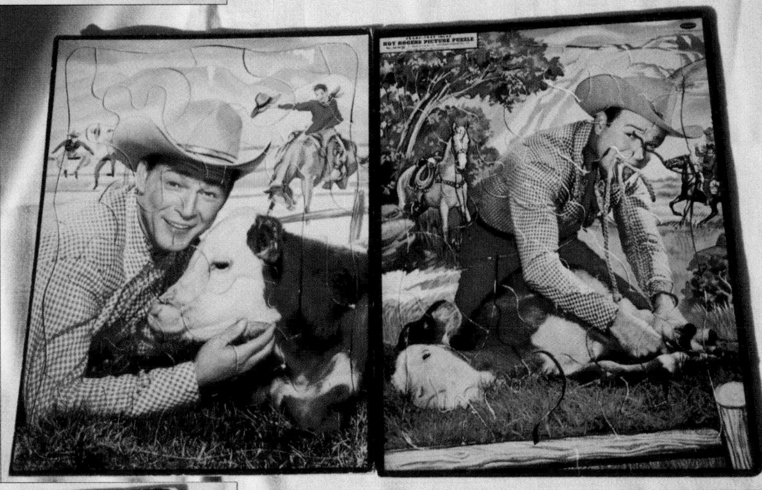

Roy Rogers frame tray inlay 9 1/2″ x 11″ picture puzzles, 1950s. (G.S.)
#1. Roy Rogers with steer calf. Painted background. #2. Roy Rogers with calf. Painted background.
Value $35-50 each

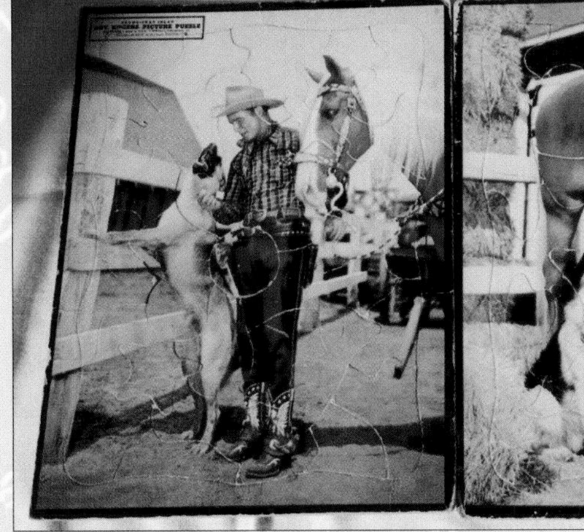

Roy Rogers frame tray inlay 11 1/2″ x 13″ picture puzzles, 1950s. (G.S.)
#1. Roy Rogers, Trigger, and Bullet.
#2. Trigger and Bullet.
Value $35-50 each

Roy Rogers frame tray inlay 9 1/2″ x 11″ picture puzzles, 1950s (G.S.)
#1. Roy Rogers with silver trimmed saddle.
#2. Roy Rogers whittling. Painted background.
Value $35-50 each

Roy Rogers frame tray inlay
11 1/2″ x 14 7/8″ picture puzzles, 1952. (G.S.)
#1. Roy Rogers putting show saddle on Trigger.
#2. Roy Rogers holding gold-plated gun.
Value $35-50 each.

Roy Rogers frame tray inlay 9 1/2″ x 11″
picture puzzles, 1950s. (G.S.)
#1. Roy Rogers holding gun and Trigger.
#2. Roy Rogers on rearing Trigger.
Value $35-50 each

Roy Rogers frame tray inlay
11 1/2″ x 14 7/8″ picture puzzles, 1950s. (G.S.)
#1. Roy Rogers and Trigger.
#2. Roy Rogers, Dusty, and Dale Evans sitting on fence.
Value $50-85 each

Roy Rogers frame tray inlay 11 1/2″ x 13″
picture puzzles. 1950s. (G.S.)
#1. Roy Rogers standing along side of Trigger.
#2. Roy Rogers in front of gate.
Value $50-85 each

Roy Rogers, Dale Evans, Bullet, and Trigger 3 1/2" x 5" inlay tray puzzles. (C.Q.)
Value $25-40 each

Gabby Hayes carrot farm large frame tray inlay puzzle. Made in 1950s by Milton Bradley Co. (G.S.)
Value $55-90

TOY BOXES

Roy Rogers toy box with brown illustrations of Roy, Bullet, and Trigger. Approximately 12" x 16" x 16". Padded vinyl top, 1950s. Rare. (B.W.)
Value $225-350

Roy Rogers toy chest. Wood with black illustrations of Roy and Trigger. Early 1950s. Rare. (G.S.)
Value $350-450

INDEX